AF342131

Springer Tracts in Modern Physics
Volume 207

Managing Editor: G. Höhler, Karlsruhe

Editors: J. Kühn, Karlsruhe
T. Müller, Karlsruhe
A. Ruckenstein, New Jersey
F. Steiner, Ulm
J. Trümper, Garching
P. Wölfle, Karlsruhe

Springer Tracts in Modern Physics

Springer Tracts in Modern Physics provides comprehensive and critical reviews of topics of current interest in physics. The following fields are emphasized: elementary particle physics, solid-state physics, complex systems, and fundamental astrophysics.

Suitable reviews of other fields can also be accepted. The editors encourage prospective authors to correspond with them in advance of submitting an article. For reviews of topics belonging to the above mentioned fields, they should address the responsible editor, otherwise the managing editor. See also springeronline.com

Managing Editor

Gerhard Höhler

Institut für Theoretische Teilchenphysik
Universität Karlsruhe
Postfach 69 80
76128 Karlsruhe, Germany
Phone: +49 (7 21) 6 08 33 75
Fax: +49 (7 21) 37 07 26
Email: gerhard.hoehler@physik.uni-karlsruhe.de
www-ttp.physik.uni-karlsruhe.de/

Elementary Particle Physics, Editors

Johann H. Kühn

Institut für Theoretische Teilchenphysik
Universität Karlsruhe
Postfach 69 80
76128 Karlsruhe, Germany
Phone: +49 (7 21) 6 08 33 72
Fax: +49 (7 21) 37 07 26
Email: johann.kuehn@physik.uni-karlsruhe.de
www-ttp.physik.uni-karlsruhe.de/~jk

Thomas Müller

Institut für Experimentelle Kernphysik
Fakultät für Physik
Universität Karlsruhe
Postfach 69 80
76128 Karlsruhe, Germany
Phone: +49 (7 21) 6 08 35 24
Fax: +49 (7 21) 6 07 26 21
Email: thomas.muller@physik.uni-karlsruhe.de
www-ekp.physik.uni-karlsruhe.de

Fundamental Astrophysics, Editor

Joachim Trümper

Max-Planck-Institut für Extraterrestrische Physik
Postfach 13 12
85741 Garching, Germany
Phone: +49 (89) 30 00 35 59
Fax: +49 (89) 30 00 33 15
Email: jtrumper@mpe.mpg.de
www.mpe-garching.mpg.de/index.html

Solid-State Physics, Editors

Andrei Ruckenstein
Editor for The Americas

Department of Physics and Astronomy
Rutgers, The State University of New Jersey
136 Frelinghuysen Road
Piscataway, NJ 08854-8019, USA
Phone: +1 (732) 445 43 29
Fax: +1 (732) 445-43 43
Email: andreir@physics.rutgers.edu
www.physics.rutgers.edu/people/pips/
Ruckenstein.html

Peter Wölfle

Institut für Theorie der Kondensierten Materie
Universität Karlsruhe
Postfach 69 80
76128 Karlsruhe, Germany
Phone: +49 (7 21) 6 08 35 90
Fax: +49 (7 21) 6 08 77 79
Email: woelfle@tkm.physik.uni-karlsruhe.de
www-tkm.physik.uni-karlsruhe.de

Complex Systems, Editor

Frank Steiner

Abteilung Theoretische Physik
Universität Ulm
Albert-Einstein-Allee 11
89069 Ulm, Germany
Phone: +49 (7 31) 5 02 29 10
Fax: +49 (7 31) 5 02 29 24
Email: frank.steiner@physik.uni-ulm.de
www.physik.uni-ulm.de/theo/qc/group.html

Innocent Mutabazi   José Eduardo Wesfreid
Etienne Guyon

Editors

Dynamics of Spatio-Temporal Cellular Structures

Henri Bénard Centenary Review

With 115 Figures

 Springer

Innocent Mutabazi
Laboratoire de Mécanique
  Physique et Géosciences
Université du Havre
F-76058 Le Havre Cedex
France
mutabazi@univ-lehavre.fr

José Eduardo Wesfreid
Physique et Mécanique
  des Milieux Hétérogènes
Ecole Supérieure de Physique
  et Chimie Industrielles de Paris
F-75231 Paris Cedex 05
France
wesfreid@pmmh.espci.fr

Etienne Guyon
Physique et Mécanique
  des Milieux Hétérogènes
Ecole Supérieure de Physique
  et Chimie Industrielles de Paris
F-75231 Paris Cedex 05
France
guyon@pmmh.espci.fr

Physics and Astronomy Classification Scheme (PACS): 47.25, 47.10, 44.60, 05.70Ce

Library of Congress Cataloging-in-Publication Data
Dynamics of spatio-temporal cellular structures/edited by Innocent Mutabazi, José
  Eduardo Wesfreid, Etienne Guyon.
      p. cm. — (Springer tracts in modern physics, ISSN 0081-3869; v. 207)
   Includes bibliographical references and index.
   ISBN 0-387-40098-2 (alk. paper)
    1. Bénard cells.   2. Heat—Convection.   3. Kinetic theory of gases.   4. Bénard, Henri,
   1874–1939.   5. Marangoni effect.   6. Nonlinear theories.   7. Chaotic behavior in systems.
   I. Mutabazi, Innocent.   II. Wesfreid, J.E.   III. Guyon, Etienne.   IV. Springer tracts in
   modern physics; 207.
   QC1.S797 Bd. 207
   [QC330.2]
   536′.25—dc22   2005043231

ISBN-10: 0-387-40098-2   e-ISBN 0-387-25111-1
ISBN-13: 978-0387-40098-3

Printed on acid-free paper.

springeronline.com

Preface

On March 15, 1901, Henri Bénard defended his thesis entitled "Les Tourbillons cellulaires dans une nappe liquide propageant de la chaleur par convection en régime permanent"[1] at the University of Paris, Sorbonne. The results contained in this thesis have been at the origin of recent intensive research activities on cellular structures observed in many physicochemical systems far from equilibrium: instabilities, spatio-temporal patterns, chaos, and turbulence.

The French Physical Society organized a scientific meeting to commemorate the centenary of Bénard's thesis, at the Ecole Supérieure de Physique et Chimie Industrielles de Paris (ESPCI). This meeting, which gathered approximately one hundred scientists and graduate students working in nonlinear science, was honored by the presence of the director of the ESPCI, Professor Pierre-Gilles de Gennes, Nobel laureate in physics (1991), who gave the opening talk.

At the conference, lectures were given by internationally recognized scholars who have contributed to the development of Bénard's work: J.E. Wesfreid, P. Manneville, Y. Pomeau, M. Velarde, J. Gollub, M. Provansal, G. Nicolis, B. Castaing, and P. Coullet. A poster session and a round table on further developments in nonlinear physics were organized.

In the present book, we have extended the list of contributors in order to cover all the aspects involved with Bénard's work, with a main focus on thermal convection, on Bénard–Marangoni instability and on Bénard–von Karman instability.

We would like to thank Dr. Hans Koelsch from Springer for the publication of this monography in the Springer Tracts in Modern Physics series. We acknowledge a critical reading by C.D. Mitescu and a very helpful technical assistance from Olivier Crumeyrolle.

Le Havre, Paris
10 May 2004

Innocent Mutabazi
José Eduardo Wesfreid
Etienne Guyon

---

[1] Cellular vortices in a thin liquid layer propagating heat by convection in a stationary regime.

Contents

Part IV   Bénard and Wakes

Part V   Extension of Bénard's Work

Contributors

G. Ahlers, Department of Physics and IQUEST, University of California, Santa Barabara, CA 93106, USA, guenter@physics.ucsb.edu

F.H. Busse, Institute of Physics, University of Bayreuth, D-95440 Bayreuth, Germany, busse@univ-bayreuth.de

A. Chiffaudel, Groupe Instabilités et Turbulence, Service de Physique de l'Etat Condensé, DRECAM, CNRS URA 2464, CEA-Saclay, F-91191 Gifsur-Yvette Cedex, France, arnaud.chiffaudel@cea.fr

F. Daviaud, Groupe Instabilités et Turbulence, Service de Physique de l'Etat Condensé, DRECAM, CNRS URA 2464, CEA-Saclay, F-91191 Gifsur-Yvette Cedex, France, daviaud@spec.saclay.cea.fr

O. Dauchot, Groupe Instabilités et Turbulence, Service de Physique de l'Etat Condensé, DRECAM, CNRS URA 2464, CEA-Saclay, F-91191 Gifsur-Yvette Cedex, France, dauchot@amoco.saclay.cea.fr

B. Dubrulle, Groupe Instabilités et Turbulence, Service de Physique de l'Etat Condensé, DRECAM, CNRS URA 2464, CEA-Saclay, F-91191 Gifsur-Yvette Cedex, France, bdubru@spec.saclay.cea.fr

K. Eckert, Institute of Aerospace Engineering, Dresden University of Technology, D-01062 Dresden, Germany, eckert@tfd.mw-tu-dresden.de

N. Garnier, Laboratoire de Physique, CNRS UMR 5672, ENS-Lyon, 46 alle d'Italie, F-69264 Lyon Cedex 07, France, nicolas.garnier@ens-lyon.fr

J. Gollub, Physics Department, Haverford College, Haverford, PA 19041, and Department of Physics and Astronomy, University of Pensylvania, Philadelphia, PA19104, USA, jgollub@haverford.edu

S. Goujon-Durand, Université Paris XII, 61 Avenue du Général de Gaulle, F-94010 Créteil, France, sophie@pmmh.espci.fr

E. Guyon, Physique et Mécanique des Milieux Hétérogènes, Ecole Supérieure de Physique et Chimie Industrielles de Paris, 10 rue Vauquelin, F-75231 Paris Cedex 05, France, guyon@pmmh.espci.fr

P. Manneville, Laboratoire d'Hydrodynamique, Ecole Polytechnique, F-91128 Palaiseau Cedex, France, paul.manneville@ladhyx.polytechnique.fr

I. Mutabazi, Laboratoire de Mécanique, Physique et Géosciences, Université du Havre, 25, rue Philippe Lebon, F-76058 Le Havre Cedex, France, mutabazi@univ-lehavre.fr

Y. Pomeau, Laboratoire de Physique Statistique de l'Ecole Normale Supérieure, 24, rue Lhomond, F-75231 Paris Cedex 05, France, pomeau@physique.ens.fr

A. Prigent, Laboratoire de Mécanique, Physique et Géosciences, Université du Havre, 25, rue Philippe Lebon, F-76058 Le Havre Cedex, France, arnaud.prigent @univ-lehavre.fr

M. Provansal, Institut de Recherche sur les Phénomènes Hors Equilibre, UMR CNRS 6594, Technopole Chateau-Gombert, 49, rue F. Joliot-Curie, B.P. 146, F-13384 Marseille Cedex 13, France, michel.provansal@univ-mrs.fr

A. Thess, Department of Mechanical Engineering, Ilmenau University of Technology, P.O. Box 1000565, D-98684 Ilmenau, Germany, thess@tu-ilmenau.de

M. Velarde, Instituto Pluridisciplinar, Universidad Complutense de Madrid, Paseo Juan XXIII, n1, 28040 Madrid, Spain, and International Center of Mechanical Sciences (CSIM), Palazzo del Torso, Piazza Garibaldi, 33100 Udine, Italy, velarde@fluidos.pluri.ucm.es

J.E. Wesfreid, Physique et Mécanique des Milieux Hétérogènes, Ecole Supérieure de Physique et Chimie Industrielles de Paris, 10 rue Vauquelin, F-75231 Paris Cedex 05, France, wesfreid@pmmh.espci.fr

Part I

Introduction

1 The Context of Bénard Scientific Work and Nonlinear Science

Innocent Mutabazi[1], José Eduardo Wesfreid[2], and Etienne Guyon[2]

[1] Laboratoire de Mécanique, Physique et Géosciences, Université du Havre,
25, rue Philippe Lebon, BP.540, 76058 Le Havre Cedex, France
[2] Physique et Mécanique des Milieux Hétérogènes(PMMH –UMR 7636 CNRS),
Ecole Supérieure de Physique et de Chimie Industrielles de Paris,
10, rue Vauquelin, 75231 Paris Cedex 05, France

The last thirty years have seen intense development of research activites in cellular structures in physicochemical systems. Although the work of Henri Bénard received rather little recognition for a long period, particularly in France, it became seminal in the 1970's. Beginning with the experiments of Pierre Bergé and Monique Dubois in Saclay and theoretical support from Pierre Gilles de Gennes and Yves Pomeau, French researchers displayed a renewed interest in the thermal convection problem. New teams in Paris (A. Libchaber in the Ecole Normale Suprieure known as "ENS"), Orsay (Orsay group on liquid crystals), and other research centers all over the country started to work intensively on this subject and related topics. In different countries (Belgium, Germany, the Soviet Union, USA, Spain, Italy), Bénard convection became a table-top physical system that gave rise to many important results using theoretical, numerical, and experimental tools. In fact, Bénard convection stimulated the development of new tools of investigation as Laser light scattering and Laser Doppler anemometry, chaotic signal processing, spatio–temporal diagrams technique, and so on.

Today, the Bénard problem and, more generally, the study of thermoconvective instabilities , has awakened the interest of physicists in fluid mechanics problems thanks to a multiple conjunction: in the 1970's there was a considerable interest in global theories of phase transitions and critical phenomena, and Bénard convection offered a simple example of a mean field-like transition around a convection threshold (e.g., Ginzburg–Landau theory: coherence length, critical exponents, ...). On the other hand, the approach of the Brussels school of statistical physics under the influence of Prigogine connected dynamical convective states to "far from equilibrium" thermodynamics and looked for the applications of some general criteria. Finally, the work of Ruelle and Takens came as a "bomb" to focus this interest on thresholds of noisy states in dissipative systems. Physicists such as P. Martin at Harvard on the theoretical level, and G. Ahlers at Bell Labs and A. Libchaber in Paris, on the experimental side, soon came up with new results although they generally dealt with extended (large aspect ratio) boxes having a large number of degrees of freedom as opposed to small aspect ratio boxes. The interest in dissipative crystallography led to the production of spectacular images of convection patterns reproducing, in particular, some nice illustrations originally due to H. Bénard. A few years later, these very same scientists also considered the presence and organization of defects and textures, a field that is still active at the present time. The existence of dynamical order far from equilibrium also triggered interest in other communities, in particular in the life sciences, concerned with the origin and organization of life since the large initial disordered "soup" (although this does not appear today to be the major basic mechanism). All these subjects referred to the Rayleigh–Bénard problem, the "fruit fly" equivalent for dissipative systems! It should be mentioned however, for completeness, that these treatments did not get very deep into the mechanical aspect of the problems nor have much to do with researchers working in thermal physics, for example.

The second major contribution of Bénard is the investigation of the wake behind bluff bodies, and his name should rightly be associated with that of Theodore von Karman (1912), as the first of Bénard's papers was published four years before that date. The double array of alternating rolls—the vortex street—is an example of a coherent structure that survives even at high Reynolds numbers, although this aspect has been studied far less than that of the mixing layers. Similar treatments of phase transitions also apply to the case of the Bénard–von Karman structures. Finally, there are many present developments of the Bénard–von Karman problem with major practical and industrial applications in particular those dealing with engineering tools (as the vortex-induced vibrations problem in fluid structure interaction), with passive and active control, and with the connection with turbulent flows.

Thus, Bénard has investigated two major problems that have played a leading role in the development of hydrodynamic stability theory, and which are formulated today in terms of *closed systems* (thermal convection) and *open-flow systems* (wakes). It is now acknowledged that there exist major differences in the transition to chaos and turbulence between such closed and open-flow systems. The impact of Bénard's work through these two major discoveries, as well as the beauty of the experiments Bénard developed to characterize them, well deserve the recognition provided by the present publication.

In a famous series of lectures given at the University of Baltimore in 1884 and recently reedited by MIT Press (1987), Lord Kelvin wrote "it seems to me that the test of *do we not understand* or *do we understand a particular point of physics? is can we make a mechanical model of it?*". The work and heritage of Henri Bénard have dealt largely with this view of the unity of physics, which was well recognized by scientists at the end of the nineteenth century but was sometimes forgotten later, in particular after the end of the Second World War. The commemoration of the centenary of Bénard's thesis inevitably raises, within and beyond his work, the question of the relationship between physics and mechanics. Bénard's scientific productivity is characterized by a remarkable richness in the observations and images that used all the available physical techniques of his time. He tracked the structures and 3-D profiles of hexagonal cells as well as the arrays of vortices behind obstacles of various shapes, with experimental talent and imagination, especially thanks to the use of the then-developing cinematography. This general theme of visualization and instrumentation has remained a subject of general interest in fluid mechanics. Not only has the laser revitalized and refined the acquisition and treatment of images and data (e.g., laser tomography; PIV), but also the use of elastic- and inelastic-scattering spectroscopy has allowed access to diffusive or convective variables in flow fields. All the various techniques of physics experimentation (its tool box) have been, or will be, used at some stage for the study of flows, although the ignorance of fluid mechanics by many physicists has sometimes delayed this use. In France, a few decades ago, the situation was such that a student graduating in physics could ignore the concept of "viscosity" or "boundary layer". In fact, the heavy mathematization of mechanics teaching as well as, paradoxically, its strong engineering orientation, did not favor a simple introduction of continuum mechanics in the basic training of physicists in France. The origin of this situation can be traced back to the nineteenth century when France produced a series of great mathematicians who were also engineers, such as Navier who produced the first formulation of the Navier-Stokes equations in 1823. Navier was also an engineer who graduated from the "Ecole des Ponts et Chaussées". However, throughout his life, he strongly resented the fact that having calculated very precisely and tightly a suspension bridge in Paris, improving through his theoretical calculations the empirical production of British engineers, his bridge had to be destroyed before being actually used, for lack of a large enough safety factor. The grand écart is not an easy exercise!

Bénard's work also raises the question of the relationship between theory and experiment. It is interesting to note that Bénard's work was not broadly known prior to this 1970's explosion, although a limited community had continued to work on the subject, without major new discoveries. One may note that, in France, no major theoretical effort accompanied Bénard's work, which was seen more as a curiosity (a referee comment already noted on his thesis report). This is different from what was taking place in the United Kingdom where the work of Lord Rayleigh and G.I. Taylor provided explanations for the observed phenomena, and led to experiments well coupled with a theoretical approach. The absence of a similar approach in France probably explains the limited impact of Bénard's work. The genius of Taylor, in particular, who explored so many problems in physics even outside fluid mechanics was well popularized in France practically at the same time as the developments of the 1970 critical phenomena studies, in particular thanks to the development of soft-matter physics. This spirit of Taylor was well present in an Advanced Studies Institute that took place in Les Houches in 1973, where some of the best experts in Mechanics of the time were the lecturers (P. Germain, S. Orszag, E. Siggia, K. Moffatt) whereas some already very senior physicists (such as P.G. de Gennes) were students! De Gennes subsequent lecture series at the Collège de France and in Seville, in the following year, made use of this school's training while giving it his own style, and produced a snowball effect within the physics community. Recent international meetings in mechanics or physics clearly show that the gap between mechanics and physics is getting smaller for the young generation of researchers, in particular in Bénard's own country. One would hope that the present reorganization of university curricula in Europe will include recommendations on their context such that, in particular, a suitable offer of joint advanced training in physics and mechanics be presented to students in engineering and research programs.

In addition to his scientific work, Bénard was also interested in the promotion of the physical sciences around the country. H. Bénard was president of the French Physical Society in 1928, succeeding Paul Langevin (1926) and Louis Lumière (1927) and before Jean Perrin (1929). P. Langevin and J. Perrin were better known for their contribution in Physics and L. Lumière for the invention of cinematography. Also note that the great mathematical physicist Henri Poincaré, who considered himself primarily a physicist, had been the president in 1902. In the context of the diffusion of science, it should be noted that at the creation, in 1937, of the Palais de la Découverte, the first museum devoted to the presentation of active (hands-on) science, H. Bénard indeed presented convection experiments in the meteorology section in connection with wind-generated cloud patterns.

The present book, containing the work of some of the major contributors in the field,is an excellent example of the richness that can result from joint studies of similar problems using the full set of experimental, theoretical, and numerical tools developed in the physical sciences. In the spirit of the conference, the book covers all the aspects involved in Bénard's work, with a main focus on thermal convection, on the Bénard–Marangoni instability, and on the Bénard–von Kármán instability. The book is made up of three parts.

Part I contains the scientific biography of H. Bénard. This historical review has been written by one of us (J.E.W.) after having collected many documents from the archives of different institutions with which H. Bénard had been associated. We found it instructive to include this extended paper on Bénard's scientific biography which shows an atypical evolution of a scientist of the twentieth century, and the place occupied by some branches of classical physics in that period dominated by the development of modern physics (quantum and subatomic physics, condensed-matter physics, and general relativity).

Part II contains articles by P. Manneville, Y. Pomeau, G. Ahlers and F. Busse. The first two, theoreticians of the Saclay group, have produced some major contributions in the study of the amplitude equations (to which the names of Newell, Whitehead, Segel, Swift, and Hohenberg should also be added for hydrodynamic instabilities) and the chaotic regime of intermittence. The third physicist provided some of the most careful experiments by making use of tools imported from other fields of condensed-matter physics and the study of critical phenomena. The contribution of F. Busse gives an application of Bénard's problem to geophysics (earth mantle, cloud structures, sunspots). It makes use of different important developments on nonlinear problems by the author in hydrodynamics and their extension to geophysical problems.

Part III addresses the problem of Bénard–Marangoni cellular structures. As the original Bénard experiment was performed with a free upper surface and involved the effect of the change of surface tension with temperature (Marangoni effect), M. Velarde, who has produced much work on the theoretical development of Bénard–Marangoni instability, presents a new approach of predictions of overstability and onset of transverse and longitudinal waves in a Bénard layer heated from above. The contribution of K. Eckert and A. Thess contains recent experimental results on secondary instability modes in Bénard convection. A recent experiment on hydrothermal waves related to Marangoni convection, described by N. Garnier, A. Chiffaudel, and F. Daviaud, is an example of the extension of the Saclay group activity.

Although experiments on wakes had occupied a large part of Bénard's lifetime, our conference devoted only a smaller part on vortex shedding in wakes, which is, however, a subject of major current interest in fluid mechanics. The contribution by M. Provansal thoroughly presents the properties of wakes behind bodies of different shapes (cylinders, spheres) and the application of Landau theory to this experiment which was developed extensively in Marseilles. The problem of spatial inhomogeneity in convection and wakes (amplitude envelope) is addressed in the chapter of S. Goujon-Durand and J.E. Wesfreid.

The last part is concerned with two topics related to Bénard's problem. J. Gollub gives here a description of patterns and quasistructures in Faraday surface waves. The last chapter presents the Couette–Taylor system which can be considered as the hydrodynamic twin of the Bénard problem. Indeed, the history arising from the experiments of Maurice Couette and Arnulph Mallock could be paralleled with that dealing with the Bénard instability in relation to the works of Rayleigh and Taylor. Taylor–Couette instability has also led to a very significant progress in the understanding of hydrodynamic stability and turbulence in closed systems.

Readers will find a list of references at the end of each contribution. Here we have just suggested a choice of a few books that thoroughly address the subjects of the Rayleigh–Bénard, Bénard–Marangoni, and Bénard–von Karman instabilities.

References

1. S. Chandrasekhar, *Hydrodynamic and Hydromagnetic stability*, Oxford University Press, London (1961).
2. G.Z. Gershuni and E.M. Zhukhovitskii, *Convective Stability of Incompressible Fluids*, Israël Program for Scientific Translations, Jerusalem (1976).
3. D.D. Joseph, *Stability of Fluid Motions II*, Springer Tracts in Nonlinear Philosophy vol 28 (1976).
4. P.G. Drazin and W.H. Reid, *Hydrodynamic Stability*, Cambridge University Press, New York (1981).
5. P. Bergé, Y. Pomeau and Ch. Vidal, *L'ordre dans le Chaos*, Herman, Paris (1984). An English translation is available under the title *"Order within Chaos"*, Wiley, New York (1984).
6. J.E. Wesfreid and S. Zaleski (eds.), *Cellular Structures in Instabilities*, Springer-Verlag Berlin Germany (1984).
7. J.K. Platten and J.C. Legros, *Convection in Liquids*, Springer-Verlag, New York (1984).
8. P. Manneville, *Dissipative Structures and Weak Turbulence*, Academic Press, New York (1991).

9. E.L. Koschmieder, *Bénard Cells and Taylor Vortices*, Cambridge University Press, New York (1993).
10. M.M. Zdravkovich, *Flow around Circular Cylinders*, Vol 1 & 2, Oxford Science Publications, Oxford, UK (1997).
11. A.V. Getling, *Rayleigh-Bénard Convection: Structures and Dynamics*, Advanced Series in Nonlinear Dynamics, World Scientific, Singapore (1998).
12. E. Guyon, J.P. Hulin, L. Petit and C.D. Mitescu, *Physical Hydrodynamics,* Oxford (2001). This is an english translation of the french second edition of the book "Hydrodynamique Physique by E. Guyon, J.P. Hulin and L. Petit.
13. P. Colinet, J.C. Legros and M.G. Velarde, *Nonlinear Dynamics of Surface Tension Driven instabilities*, Wiley-VCH, New York (2001).
14. A.A. Nepomnyanshchy, M.G. Velarde and P. Colinet, *Interfacial Phenomena and Convection*, Chapman & Hall-CRC, London (2002).

2 Scientific Biography of Henri Bénard (1874–1939)

José Eduardo Wesfreid

Physique et Mécanique des Milieux Hétérogènes (PMMH –UMR 7636 CNRS)
Ecole Supérieure de Physique et Chimie Industrielles de Paris (ESPCI)
10, rue Vauquelin, 75005 Paris, France
email address: wesfreid@espci.fr

2.1 Biographical Notes

Henri Claude Bénard was born at Lieurey, a small French village in the region of Eure, in Normandy, on October 25th, 1874. He was the only son of Felix A. Bénard (1851-1884) and Hélène M. Mangeant (1837-1901) [1-3]. His father was a small investor, who died very young. H. Bénard finished elementary school in the district of Lisieux and in Caen, nearby his birthplace, and moved to Paris to continue his studies at the Lycée Louis le Grand, one of the best high schools in France. In 1894, he succeeded in the highly competitive entrance examinations to the prestigious Ecole Normale Supérieure in Paris[1]. Indeed, this year, 17 students were selected from 307 candidates in the sciences section and 25 from 205 candidates in the humanities section [4].

H. Bénard studied with some subsequently very well-known companions (the *Normaliens*): he was a classmate of the physicist Paul Langevin and of the mathematician Henri Lebesgue[2]. In humanities studies, the 1894's promotion included Charles Péguy (poet), Albert Mathiez (historian), and Léon Bloch (who started in Literature but worked as a physicist, in collaboration with his brother Eugène). They participated in the activities of the centenary of the ENS and knew the first manifestations of the intelligentsia in favor of the Captain Dreyfus affair initiated by Paul Dupuy and Lucien Herr, the librarian of the ENS.

In 1897, Henri Bénard obtained the degree of *agrégé de physique* and began to work, in the chair of experimental physics at the Collège de France[3], as the

---

[1] The Ecole Normale Supérieure (also known as "ENS" or Ulm, from the name of the street where it is located) is a French Grande Ecole founded during the French Revolution by a decree of the Convention. Originally meant to train high school teachers, it became an elite institution, training researchers, university professors, and civil servants. It focuses on training through research, with an emphasis on the freedom of curriculum [5].

[2] Paul Dupuy, the main supervisor of the ENS wrote [6]: *"in 1894, I see Paul Langevin getting to the head of a promotion of the Ecole Normale, rich in promises, along with Henri Béghin, Henri Bénard, Noël Bernard, Henri Lebesgue and Paul Montel"*. Indeed, this promotion gave four members to the Institute (the Academy of Sciences of Paris).

[3] The Collège de France is an institution dedicated to knowledge, created in 1529 by the King François I. It has always been independent of any university and free from

assistant of Eleuthère Mascart and Marcel Brillouin[4]. During the first year with Mascart, a specialist in optics, he studied the angle through which polarized light is rotated by sugar in solution. But it was during his second year at the Collège de France that he began to be interested in fluid dynamics. Actually, Marcel Brillouin wanted to repeat Poiseuille's experiments on water-flow-rate laws. He carried out these experiments with mercury, in order to study the influence of viscosity. Great experimental skill was required, especially when measuring the diameter and the cross section of capillary tubes used to study the flux of mercury. Marcel Brillouin wrote, some years later, lectures titled *Lessons about the Viscosity of Liquids and Gases* in which he particularly referred to Bénard's experiments on viscosity in mercury [9]. Under Brillouin's direction, Bénard did the French translation of the second volume of Boltzmann's *Lectures on Gas Theory*, published in 1905. At the same time, Bénard was preparing his Ph.D. thesis. Observing by chance the motion of graphite particles in a molten paraffin bath, he became interested in the organization displayed by particles on the bottom layer of the liquid. Using optical methods, he then studied the movement of the particles in a layer of fluid heated from below, paying special attention to the deformation of the free surface due to convection, using the knowledge and experience he had acquired with Mascart and Brillouin. On March 15, 1901, he defended his thesis before a committee composed of Gabriel Lippmann (1845-1921, Nobel Prize in Physics in 1908), Edmond Bouty (1846-1922) and Emile Duclaux (1840-1904). The second subject of the thesis[5] dealt with the rotation of plane-polarized light by sugar in solutions. The jury did not place enough value on the consequences and meaning of this Ph.D. thesis. In his report on Bénard's thesis, previous to the defense, E. Bouty said that the subject was innovative, and the thesis, a very good one, much beyond the average of the other theses, its main interest lying in the application of a wide range of optical methods, but at the same time he pointed out that Bénard did not make any effort to provide general theoretical explanations to the laws found through experiments. But the

---

supervision. The lectures are open to the public without registration. The Collège de France does not deliver any certificate or degree. Nowadays, its range of studies includes humanistic and scientific fields. Its faculty staff includes many distinguished scholars.

[4] E. Mascart (1837-1908), Professor of Experimental Physics at the Collège de France from 1872 to 1908, was one of the first to study the influence of the earth on optic phenomena. He introduced Maxwell's electromagnetism in France and became very well known for his work on electricity [7]. M. Brillouin (1854-1948) was Professor of Mathematical Physics at the Collège de France from 1900 to 1931. He wrote over 200 experimental and theoretical papers on topics including the kinetic theory of gases, viscosity, thermodynamics, electricity, and physics of condensed matter. Precursor of wave mechanics, and an open-minded scientist, he extended his work from the history of science to the physics of the earth and of the atom. Both, Mascart and Brillouin, were respectively, grand-father and father of another professor of the Collège de France, the physicist Leon Brillouin (1889-1969) [8].

[5] A second subject, mostly bibliographical, was at that time, required by the jury at the final stage of doctoral studies.

Fig. 2.1. The 1894 sciences promotion at the Ecole Normale Supérieure, in 1896. Seated, from left to right: Renaud, Massoulier, Béghin, Langevin, **Bénard** and Montel. Standing, from left to right: Lebesgue, Bernard, Foulon, Angelloz-Pessey, Patte, Cambefort, Meynier and Dubreuil (*Centre de Ressources Historiques de l'ESPCI*).

The members of the 1894 sciences promotion were: Joseph Angelloz-Pessey (Professor of Mathematics at the Lycée Buffon, ? -1932), Henri Béghin (professor at the Sorbonne, member of the Academy of Sciences; 1876-1969), Henri Bénard, Noël Bernard (professor at the University of Poitiers; 1874-1911), Georges Cambefort (? -1964), Louis Dubreuil (chemist, professor at the Collège Chaptal; 1873 -1922), Georges Foulon (? -1958), Paul Langevin (he was the eldest of the promotion, as he previously graduated from the Ecole Supérieure de Physique et Chimie Industrielles de Paris (ESPCI), whose director he became in 1925, professor at the Collège de France, and member of the Academy of Sciences; 1872-1946), Henri Lebesgue (Professor of the Collège de France, member of the Academy of Sciences; 1875-1941), Pierre Massoulier (General Inspector of high school teaching in Physics and Chemistry; 1874-1961), Paul Montel (mathematician, member of the Academy of Sciences; 1876-1975), François Meynier, Lucien Patte (professor of physics at the Lycée Charlemagne) and Jules Renaud (? -1951).

SÉRIE A N° 387,
N° D'ordre
1057.

THÈSES

PRÉSENTÉES

A LA FACULTÉ DES SCIENCES DE PARIS

POUR OBTENIR

LE GRADE DE DOCTEUR ÈS SCIENCES PHYSIQUES,

PAR

M. Henri BÉNARD.

Ancien Préparateur au Collège de France.

1ʳᵉ THÈSE. — LES TOURBILLONS CELLULAIRES DANS UNE NAPPE LIQUIDE
PROPAGEANT DE LA CHALEUR PAR CONVECTION, EN RÉGIME
PERMANENT.

2ᵉ THÈSE. — PROPOSITIONS DONNÉES PAR LA FACULTÉ.

Soutenues le   mars 1901, devant la Commission d'Examen.

MM. LIPPMANN, *Président,*
BOUTY,
DUCLAUX, *Examinateurs.*

PARIS,

GAUTHIER-VILLARS, IMPRIMEUR-LIBRAIRE
DU BUREAU DES LONGITUDES, DE L'ÉCOLE POLYTECHNIQUE
Quai des Grands-Augustins, 55.

1901

Fig. 2.2. H. Bénard's Ph.D. thesis, published by Gauthier-Villars.

report, established after the defense, mentioned that "... *though Bénard's main thesis was very peculiar, it did not bring significant elements to our knowledge. The jury considered that the thesis should not be considered as the best of what Bénard could produce*"[6].

Once H. Bénard finished his thesis, he settled down in Paris, with a fellowship from the Foundation Thiers, and got married the same year, on December 23, to Clémentine Malhèvre (1876-1943), a few months after his mother's death. They had no children. The following year he was appointed assistant professor at the Faculty of Sciences in Lyon, where he was in charge of introductory courses. At that moment, a new experimental activity began: the observation of the fluid motion when a prismatic body is moved across a container filled with liquid. Bénard was astonished by the deformations he observed on the free surface of the liquid, which he associated with the presence of vortices in the fluid. This observation led him to build an experimental facility in the university building's basement, in order to observe the deformation of the free surface of the liquid when vortex shedding occurred.

Bénard's scientific research is marked by a specific element: the use of cinematography as an instrument of observation and measurement. In fact, in the experiment performed in the Faculty of Science in Lyon, he used the movie camera as a means of observation. He published the description of these alternating vortices in two articles in the *Comptes-Rendus Hebdomadaires des Séances de l'Académie des Sciences de Paris*, in 1908 [B12, B13]. In 1910, he was appointed professor of physics at the Faculty of Science of the University of Bordeaux, where Pierre Duhem[7] was the head of the physics laboratory [10], and carried on the analysis of the movies he had made in Lyon, particularly on the wavelength and the frequency of emission of vortices as a function of different parameters such as the velocity and the dimensions of a moving body. Simultaneously, he used a movie camera to make many movies of convection for scientific popularization. In 1914, the First World War broke out and Bénard, as a former student of the Ecole Normale Supérieure, was mobilized with the rank of officer and appointed to a military scientific commission. One of the subjects he dealt with was the improvement of the refrigeration wagons transporting meat to the front. He actually conceived new methods for the measurements of the thermal diffusivity of the wagon's walls. This work was published after the war, in 1919 [B21, B23]. Later on, he was sent to the Superior Commission of Inventions of the Ministry of War, to work on different aspects of optics. He had been interested in the

---

[6] *"Bien que la thèse principale de M. Bénard d'ailleurs fort curieuse, ne paraisse pas susceptible par ses développements ultérieurs, d'ajouter grand chose de nouveau à nos connaissances, le Jury a été unanime à estimer qu'il ne fallait pas prendre cette thèse comme la mesure définitive de ce que M. Bénard peut donner."*

[7] Pierre Duhem (1861-1916) was professor of theoretical physics at the University of Bordeaux from 1894 until his death. He achieved works of leading importance in the philosophy of science, historiography of science, and science itself. His interest in science was mainly directed to areas of mathematical physics, and especially thermodynamics.

wakes produced by submarines, and had studied the traces of ships in the sea and the advantages of using polarized light. Following his propositions, the National Navy Office built a periscope with polarized prisms in spath of Island, which was also provided to the Allies. Bénard also built panoramic glasses with cylindrical lenses, trying to enlarge the image in one direction, an invention that revealed itself useful for naval applications. Between 1917 and 1919, he participated in the Directorate of Inventions of the Ministry of War of which he later became head of the physics section under the direction of Jules-Louis Breton[8].

In 1922, Bénard moved from Bordeaux to Paris, where he was appointed as assistant professor at the Faculty of Science of Sorbonne University. Four years later, in 1926, he was named full professor and taught general physics to first-year students. In 1928, he became president of the French Physical Society for one year.

As far as experiments are concerned, he carried out new studies on alternating vortices and faced difficult moments in his career due to the lack of financial support. But finally in 1929, he took part in the new Institute of Fluid Mechanics, whose creation was the result of an important cooperation between the Sorbonne University of Paris and the Ministry of Aeronautics[9], and a year later, he was appointed professor of experimental fluid mechanics. In 1932, 19 persons worked in this institute, including 12 fellowships supported by the Air Ministry. They represented 11 percent of the researchers on the staff of the Faculty of Sciences in Paris [15]. Bénard was in charge of the experimental fluid mechanics laboratory, with enough room for a very important experiment on vortex shedding: on the first floor was a container with a bluff body that produced the emission of vortices, a phenomenon observed from two floors higher above by means of a movie camera.

In his laboratory at the Institute of Fluid Mechanics, Bénard was the advisor for several theses on different aspects of natural and forced convection and also

---

[8] Jean-Louis Breton (1872-1940), deputy of the Republican Socialist Party, was head of the War Office of Invention. The Office of Research and Inventions, established in 1922, is one of the ancestors of the present CNRS.

[9] The Ministry of Aeronautics, under the technical coordination of Albert Caquot, created the Fluid Mechanics Institute of Paris at the University of Paris, and gave strong financial support to scientific research in fluid mechanics, setting up the same year, four chairs, with eight professors [11,12]. The Institute in Paris was then placed under the direction of Henry Villat (1879-1972), who worked in mathematics and in theoretical fluid dynamics. A member of the Academy, Villat became its president in 1948 [13]. He maintained a close friendship with Bénard, as they were both from Normandy. Other laboratories were opened in Marseille, under the direction of Joseph Pérès (1890-1962), who moved in 1932 to the Institute in Paris [14], in Lille under the direction of Marie-Joseph Kampé de Fériet (1893-1982) while in Toulouse, activity developed with Charles Camichel (1871-1966). Five associated chairs were also created in Caen, Lyon, Nantes, Poitiers, and Strasbourg.

Fig. 2.3. Henri Bénard, Professor in Paris.

on vortex dynamics. Among his students and collaborators we can mention[10] the following. Dusan Avsec, from the Balkans, arrived in the laboratory in 1934 and worked in thermal convection of forced flows. He also studied electroconvection jointly with Michel Luntz. The latter worked on flow singularities. C. Woronetz finished his Ph.D. in thermal convection in 1934. H. Journaud did experiments on convection rolls until 1932, and his works was followed in 1936 by Victor Volkovisky who finished his Ph.D. thesis on longitudinal rolls in 1939. Paul Schwarz presented his Ph.D. thesis in 1937 on optical methods applied to the vortex shedding experiments. V. Romanovsky studied convection in muds and G. Sartory, convection induced by radiation. François-Joseph Bourrières, a former Bénard student from Bordeaux and professor at the Lycée Stanislas, did research in the Institute of Fluid Mechanics on fluid-structure interaction. Malterre worked on solitons and L. Denes, from Hungary, on thermal conductivity. André Fortier (?-1996) experimented on the viscosity of air and gases. In this group, Robert Fabre (1908-2002) and E. Drussy were technical assistants. At the same time, in the Institute, Lucien Malavard (1910-1990) and Lucien Romani (1909-1990), assistant and technician respectively, were working on electrical analogies of phenomena of fluid mechanics in the laboratory of Joseph Pérès, assistant professor. Another group in the Institute, directed by Adrien Foch (1887-1980), assistant professor, who designed a supersonic wind tunnel, included M. Dupuy, Lucien

---

[10] According to the information we have presently, we could not verify the full list of Bénard's collaborators and not even if the ones mentioned as such in this list, were really collaborators.

Santon who did experiments on interferometry and wind tunnels, and Charles
Chartier, who accomplished a Ph.D. thesis on flow visualisation in 1937.

Henri Bénard died on March 29, 1939, at Neuilly-sur-Seine, near Paris. A.
Foch, assistant professor, succeeded Bénard as professor in experimental fluid
mechanics[11] at the Institute of Fluid Mechanics and Yves Rocard (1903-1992)
moved from Clermont-Ferrand to Paris, in order to assume the assistantship.
After the war, Y. Rocard went to the Ecole Normale Supérieure as head of the
physics department, and worked in various fields of research such as semicon-
ductors, seismology, and radio astronomy. He was one of the main leaders of

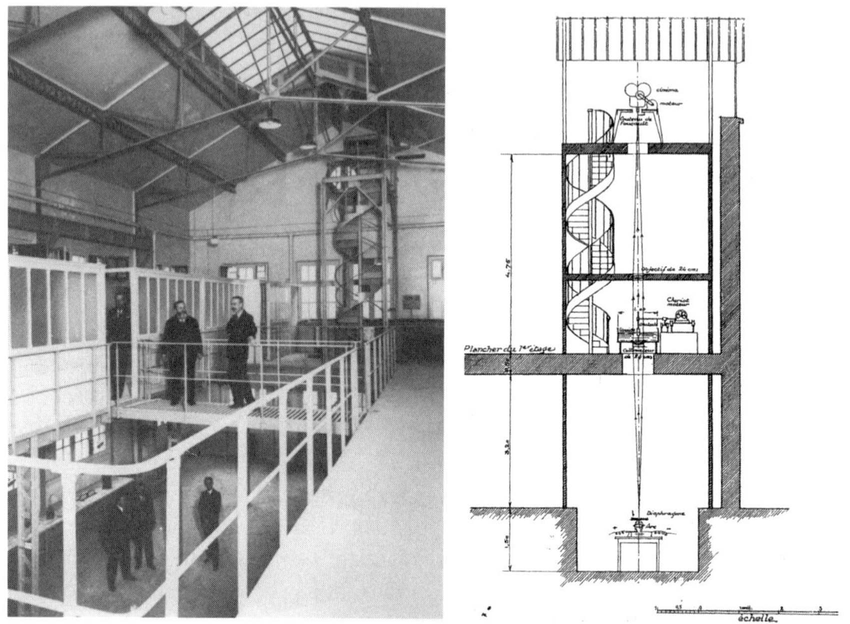

Fig. 2.4. View of the laboratory at 4, rue de la Porte d'Issy, in Paris. In the picture,
from left to right, on the upper floor, D. P. Riabouchinsky, H. Bénard and H. Villat
and on the lower floor, L. Santon, C. Woronetz, and H. Journaud. At right, scheme of
the vortex shedding installation, observed in the back of the picture.

Dimitri P. Riabouchinsky (1882-1962) an emigrant from Russia worked in the Institute. He had previously founded a private Aeronautical Institute at Koutchino (Russia)
where in 1912 he built an important wind tunnel.

the new generation of French physicists of the post-war period and reorganized
research in various fields of physics. At the same time, the Institute of Fluid
Mechanics, with H. Villat, A. Foch, and J. Pérès, originated many activities in

---

[11] Following Bénard's lectures at the Sorbonne, Léonard Rosenthal and three other
students contacted A. Foch in 1939, in order to do research on the same topics. Foch
told them that Bénard's subjects were exhausted and could no longer be subjects
for theses! (L.R., personal communication)

mechanics, especially in aeronautical and military research. In September 1946, H. Villat presided over the Sixth International Congress of Applied Mechanics in Paris, and proposed, along with Johannes M. Burgers from Delft, the creation of a permanent organization for the science of mechanics, under the form of an International Union of Theoretical and Applied Mechanics (IUTAM).

2.2 Convection Cells

In 1898, when Henri Bénard was working at the Collège de France, and trying to prepare a coherer with solid dielectrics, he observed, by chance, the presence of semiregular polygonal figures in a melted paraffin bath in which graphite dust had been incorporated. He inquired if this apparently common phenomenon had already been observed scientifically and decided to prepare laboratory experiments on thermal convection, in order to describe and measure, in a horizontal liquid layer heated from below, the convection currents that prevailed, as near as possible to their state of greatest stability. Much of his effort had been devoted to avoiding any inhomogeneity in temperature that could initiate an uncontrolled process in the movement of the liquid, a preoccupation he expressed by saying *"It is clear that, if even the littlest fluctuation or local excess of temperature, is sufficient to create a centre of ascension; how is it possible to obtain a stable regime?"* Due to the construction of an apparatus with a metal container and steam circulation, offering very homogeneous thermal conditions and a constant temperature in the lower layer of the liquid, he observed different patterns of convective movement. The main result of his observations was the discovery of a pattern of almost regular hexagonal cells, called *cellular vortices*, that is to say, a stable system with particular geometric characteristics to which he had already devoted study during his Ph.D. work. He used the expression *tourbillons cellulaires*, which were later known as Bénard cells (*cellules de Bénard.*) He insisted on the polygonal characteristics of this cellular, semi-regular vortex, due to the existence of polygons of four, five, six, and seven sides, but with a predominance of hexagons. He pointed out the difficulty of producing regular hexagons on a long surface without many defects. These cellular vortices could be generated in a steady state, under a moderate heat flux. Bénard also observed vortices in fairly volatile liquids, such as alcohol or hydrocarbon, underlying the fact that evaporation chilled the surface, causing a vertical heat flux. In order to produce a uniform thickness and to avoid evaporation problems, he worked with higher temperatures, between $50\,^\circ$ C and $100\,^\circ$ C, using substances which melted at $50\,^\circ$ C such as wax or spermaceti, a whale oil, which melts at $46\,^\circ$ C but has no significant volatility below $100\,^\circ$ C. This allowed him to create liquid films of one millimeter thickness controlled to with one micron and to obtain a spread of the thin layers, which remained constant for many hours.

Bénard studied the circulation of the liquid within the convective cell. He accurately determined the pattern of the trajectory with closed streamlines, studying the warm liquid ascending through the axis of the hexagonal cell and descending along the periphery bordered by vertical planes. He observed the

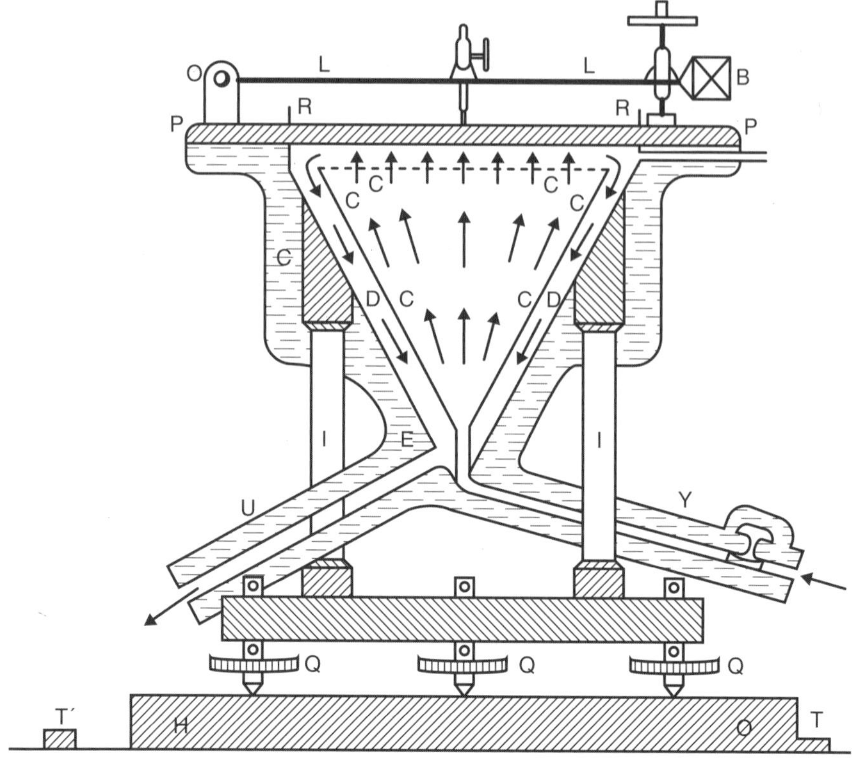

Fig. 2.5. Apparatus used by Bénard, during his Ph.D. work.

formation of a convective structure with different polygonal structures tending to a hexagonal one as a limit state, with extremely slow deformations, up to the point at which it reached a very regular shape that allowed him to measure the distance between the vertical axes of two contiguous hexagonal cells. Bénard, impressed by the periodicity discovered, studied the geometrical characteristics of these convective structures using different methods, and more precisely, the relation between the wavelength and the thickness of the liquid layer [18].

Bénard described two other types of cellular vortices that appeared in his experiments. The first one was the *tourbillons en bandes* or convective rolls, to update the expression, in a stable permanent regime whenever the heat flux is small. Later, he used Rayleigh's expression *striped-vortices* to define the phenomenon, as well as *cellular vortex of the second kind.* He tended to use the concept of *tourbillons en bande* in order to point out the fact that the structure exhibited the same spacing between strips, that is, a constant wavelength. He described a third regime called *tourbillons en chaine* or *vortex worm.* This regime, unstable and turbulent, is formed in liquids that evaporate in open air and when the heat flux is important. Cinematography was particularly useful in studying this régime. His study was mostly performed with experiments on free surfaces. Afterward, his colleague and collaborator Jean Camille Dauzère[12] became interested in experiments with an upper covered rigid surface.

One of the characteristics of Bénard's thesis work, developed in a relatively short period of time, is the enormous number of experimental methods used for the observation of the cell structures. Visualizations were achieved projecting fine licopodium grains on a free surface and drawing their trajectories on the free surface. He could also observe the border of the hexagonal cells by means of the reflecting particles of graphite powder or aluminium. As the convective

Fig. 2.6. Picture from one of the Bénard's movies, when the operator (Bénard himself?) introduces the particle tracer in the container.

motion in the liquid layer changed the pressure, the free surface was no longer flat. Therefore Bénard used the reflection produced on the deformed free surface as a concave mirror in the middle of the cells to visualize them. The best images of the polygonal structure were taken with optical transmission where the light is transmitted across the liquid layer and reflected on a steel mirror on the bottom of the recipient. He had recourse to Schlieren or Foucault's method of beam deflection on a disturbed surface to distinguish hills from valleys and to identify the rising and descending streams. This method allowed him to obtain very accurate free surface measurements. We should also note Bénard's ability

---

[12] C. Dauzère (1869-1944) was a professor of physics at the Lycée of Agen, in the south of France, and later in Toulouse. His first experiments were performed in Agen and published in 1907. During 1913-1914 he spent one year doing experiments on solidification in the industrial chemistry laboratory of Professor Charles Fabre in Toulouse and defended a Ph.D. thesis in Paris in 1919. He assumed the direction of the Pic du Midi Astronomical Observatory in 1920 until 1937, where he became interested in thunderstorms and hail [17].

to measure thickness differences of 1 micrometer for liquid layers of 1-millimeter thickness of spermaceti at an average temperature of 100 ° C. With the interference fringes obtained with a first beam reflected on the upper free surface and a second one crossing the layer, reflected on the steel mirror in the bottom of the layer, Bénard succeeded to plot the isotherms, separated by 0.1 ° C, within the convective cell.

In 1916, Lord Rayleigh (J.W. Strutt, 1842-1919) published in *Philosophical Magazine* a paper about the stability of a fluid layer subjected to a vertical gradient of temperature, referring to *"the interesting results obtained by Bénard"*. In this article, Lord Rayleigh, considering a fluid with symmetrical free - free horizontal boundary conditions, got results on the critical temperature difference necessary to produce a convective motion. From the analysis of stability, he determined the wavelength of convective cells on the instability threshold, equal to two times the thickness. The fact that this article had been published during

Fig. 2.7. Convection cells.

the war prevented Bénard from reading it before Lord Rayleigh's death, a fact he regretted because he was no longer able to discuss it with him. Several years later, in 1928 and 1934, Bénard tried to compare the predicted onset values, especially those from the new calculations from Jeffreys' work on asymmetrical rigid - free limit conditions. These conditions corresponded to the experiments he had done with a free surface. For an hexagonal structure, Bénard concluded that the theoretical value of the critical wavelength, equal to two times the thickness, was very close to the value he had obtained in his experiments. He discovered strong discrepancies between predicted and measured onset values, which were understood several years later. Myron J. Block realized experiments in 1956 [19] (similar to the ones of Bénard), and suggesting that the buoyancy could not be, in this experiment, the driving mechanism of the convection, considered that the Bénard cells are produced by variations in the surface tension with temperature. In 1958, Pearson [20] introduced the concept of surface tension gradient instability (Marangoni effect). Six years later, Nield [21] proved that the combined effect of buoyancy and Marangoni's effect caused the instability. The physical mechanism, actually responsible for the convective motion considered by Rayleigh's theoretical analysis, is the buoyancy, that is to say, the Archimedes force[13]. Bénard cited only two references as previous work on cellular convection.

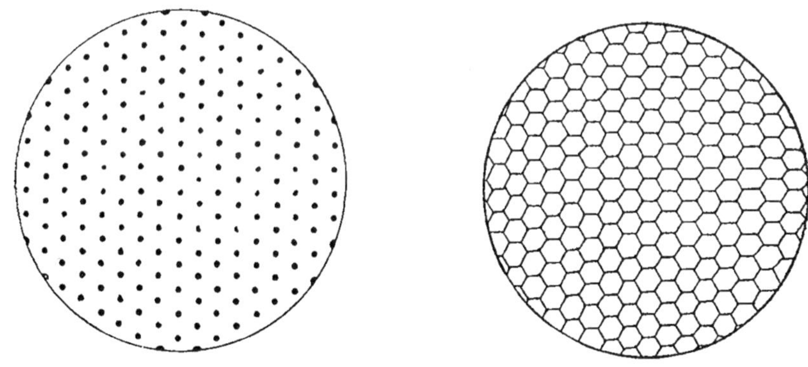

Fig. 2.8. Cells visualization by reflection and transmission: Temperature = 61,36 °C, thickness = 0,640 mm.

The first one came from E.H. Weber, who, in 1855, had described polygonal structures in drop dissolutions. Later, M.O. Lehmann gave a thermal origin of this phenomenon. He also referred to A. Guébhard, who, in 1897, observed vortex

---

[13] In fact, C. Dauzère [22] described that when he was boiling wax mixed with water, he observed a very different stability in respect to Bénard's observations. He provided an explicit explanation of this situation by the fact that the composition of the liquid modifies the surface tension and, consequently, the stability.

motion in an abandoned bath of film developer. Bénard some time later found a work by James Thomson (Lord Kelvin's brother; 1822-1892) published in 1882, entitled *On Changing Tessellated Structures in Certain Liquids*. This study dealt with the cooling of saponaceous water and the apparition of polygonal prisms.

The observation of symmetry and periodicity of the cellular vortices induced Bénard, like many other physicists of his time, to compare it with other structures observed in nature or even in life. But he knew that the thermoconvective explanation was not valid for the origin of most of the hexagonal structures observed in nature. In the thirties, he discussed the difference between living tissue and hexagonal structures, the last ones being formed by one layer of cells, and the former by several layers of superposed cells and wrongly mentioned the analogy between these latter cells and the annular cells (or Taylor-Couette rolls) observed by Geoffrey I. Taylor (1886-1975) in 1923, rolls he considered as superposed cells.

After Paul Idrac's work in 1920 [16], meteorologists became interested in the existence of convective rolls of hundreds of meters in diameter in the atmosphere[14]. The only model for understanding cloud structure was the one transverse to the wind and formed by shear layer instabilities ("Helmholtz waves" as they were called at that time). The presence of thermoconvective longitudinal rolls parallel to the wind, was a subject that Bénard promoted. He directed experiments related to this explanation, performed by his students Journaud, Avsec and Volkovisky at the Institute of Fluid Mechanics, experiments in which the fluid layer heated from below is put in motion by an inclined plane or a moving band. Dauzère observed through experiments on solidified wax that it solidifies first at the top of the border of the hexagonal convective cell because of the presence of cold currents inside the cell. From this observation and by analogy, Bénard provided a theory on the relief of craters on the moon, explained as some sort of slow cooling and solidification of surface layers of the moon minerals instead of the impact of meteorites. He also published a paper [B16] where he speculated about the analogy between cellular vortex patterns and fracture patterns in soil[15]. Bénard was aware of Jansen's observations in 1896 at the Paris Observatory about sun granulation and subscribed to a convective theory to explain these patterns. Also several scientists tried to study the polygonal shape of the granules, immediately related to turbulent convection in terms of Bénard's hexagons. Today, the role of these convective mechanisms is accepted, in the structure of solar granulation (see Chapter 6 of this book).

---

[14] Schereschewsky said that Bénard probably discussed these ideas with Idrac (1885-1935) when they worked together during the war, at the Commission of the Inventions [3].

[15] In a personal reprint of this paper, Bénard mentioning hand-script, "*Ne pas diffuser*" (not to be distributed), suggesting that he was aware of the risk of this speculation.

2.3 Vortex Shedding in Bluff Bodies

In 1898, Bénard began to be interested in experimental hydrodynamics, as he was collaborating with Marcel Brillouin at the Collège de France, on the viscosity of liquids. When he opened his laboratory at the Faculty of Science in Lyon, he began in 1904, an experiment on an obstacle, an elongated lamina, moving in the bottom of a rectangular container of 1.35 m x 0.35 m filled with a layer of 0.12 m. of water. He observed the deformation of the free surface, associated with the presence of a double trail of alternating vortices in the wake and carried out very meticulous experiments in order to obtain accurate measures of geometrical and kinematical properties. In order to study the deformation on the free liquid surface produced by pressure variation due to the existence of vortices, he used the same Schlieren optical method that had been used in thermal convection. Therefore, he was able to follow the position of the center of rotation of each vortex. Facing some technical difficulties in photographing these capillary ripples in the interface, he decided to use a movie camera, equipped with a motor, developed by Louis Lumière (1864-1948). He could then film the vortex shedding for a period of several seconds and in November 1908, he published the papers [B12,B13] in which he described two results: the existence of an alternating row of vortices and the fact that the distance between them depended only on the transverse size of the bluff body[16]. By that time, he had already analyzed more than 30,000 images from which he obtained the law of frequency as a function of velocity, as well as the geometrical characteristics of the vortex emission.

He kept on studying the results of the films from Lyon until 1925. An analysis of Bénard's laws can be found in Provansal's contribution, chapter 10 of this book. After his 1908 papers, Bénard continued to process the pictures of his experiments performed during 1908 and 1909, and in 1913 he published two new papers [B17,B18]: the improvement due to the use of the movie camera made possible the observation of the center of vortices and the precise determination of their velocity and of the distance between them. The meticulosity of the data analysis of the 133 films[17] on vortex shedding experiments that he produced, among which, after several years of research, he retained only 71, must be highlighted. From there he obtained spatio-temporal diagrams, one of which is shown in Provansal's contribution.

The analysis of this complete data set ended with four papers published in the *Comptes-Rendus de l'Académie des Sciences de Paris* in 1926 [B28-B31]. In these papers, by analyzing the period fluctuations, estimated to be between

---

[16] He later recognized that it was a speedy study and that the first one contained a drawing mistake (the sense of the vortex's rotation was inverted), mistake he corrected in 1913 [B17].

[17] During the Second World War, in June 1940, the German army held the buildings of the Institute of Fluid Mechanics in Paris. L. Malavard and L. Romani, from Pérès' laboratory, succeeded, in extremis, in taking away a rheological calculator to the Free Zone[12]. One day, the German soldiers threw in the dustbin Bénard's experimental movies on vortex shedding in order to use the cupboard in which they had been stored (R. Fabre, personal communication).

five and ten percent, Bénard performed ingenious statistical analyses of a great number of measures taken in order to obtain the law of the frequency as a function of the flow velocity. As he himself recalled, the main purpose of his thesis work on thermal convection was to measure with precision the periodicity of cellular structures. He showed the same interest when he began to study the alternating vortices behind a bluff body: his fundamental aim was to measure the frequency of emission of vortex shedding. This explains not only the use of the camera as a method of measurement, but also the great effort spent on data analysis for several years in order to obtain the laws of the variation of the frequency with the physical parameters of the experiment. The subject had become a wide topic of investigation in hydrodynamics. In Toulouse (France), C.H. Camichel, M. Teissie-Solier, L. Escande, and T. Dupin were specifically working on vortices emitted by circular cylinders. In 1928, Bénard published two other papers [B37, B38] in which he compared his observations with those of the Toulouse group, particularly on the similar properties of the law of frequency of the vortex shedding as a function of the Reynolds' number, that is to say, the nondimensional value of the fluid velocity. He expressed the frequency with the Strouhal number S[18] , which was the adequate frequency parameter he had already proposed several years earlier. Bénard did not observe a minimal Reynolds number for vortex shedding and concluded wrongly, as we now know, that there is no critical parameter for vortex shedding. In addition, he questioned the existence of similarity when comparing the results obtained with cylinders and laminas.

Lord Rayleigh stated in a paper dated 1915 [23], relating to Aeolian sound[19] and the emission of alternating vortices produced by the wind on cables, that Bénard's work had pointed out that the Strouhal's number is actually a function of the Reynolds number. Although Bénard was not the first scientist to observe vortices behind a bluff body, he may be considered as the first one to have obtained experimentally the laws characterizing their periodicity. He did not know if previous work had been done on the same phenomenon, but some years later, he discovered H.R.A. Mallock's[20] paper published in 1907 [24], to which he referred as the first work achieved on alternating vortices in fluids. He also mentioned [B41] many times the similar emission of vortices observed by Etienne Jules Marey,[21] who was certainly the pioneer of visualization tech-

---

[18] $S = fd/U$, where f is the frequency, U the fluid velocity, and d the typical size of the object.

[19] Since ancient times, it has been observed that wind causes vortex-induced vibrations of the wires of an Aeolian harp. In 1878, Strouhal found that the Aeolian tunes generated by a wire in the wind were proportional to the wind velocity divided by the thickness of the wire.

[20] Indeed, in this paper Henry Reginald Arnulph Mallock (a consulting engineer working in various branches of physics who invented and improved many instruments of high scientific value), drew different possibilities of vortex emission including the alternate raw.

[21] Étienne-Jules Marey (1830-1904) French physician, inventor, and photographer, specialized in human and animal physiology. He held the chair of Natural History of

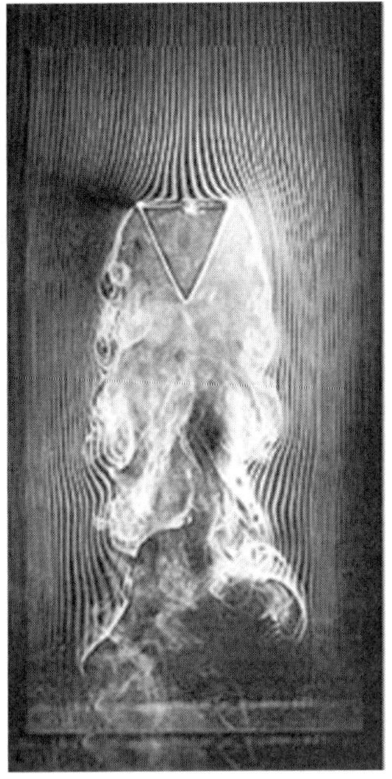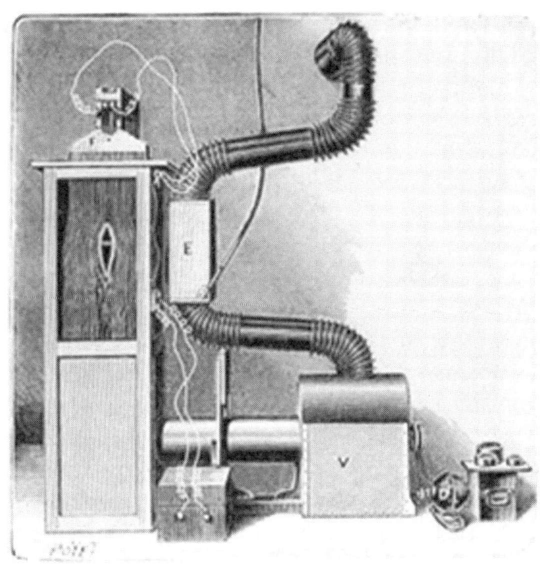

Fig. 2.9. A typical Marey's image of coherent structures (a) observed in his wind tunnel (b), published in 1901.

niques in modern fluid mechanics [25]. In 1911 and 1912 Theodor von Kármán (1881-1963) published two papers about his work on the stability of alternating vortices in a street formation achieved at the University of Gottingen. Ludwig Prandtl(1875-1953), a professor of applied mechanics, had founded a school of aerodynamics and hydrodynamics there that had acquired a world reputation, which was acquired as well by the activity in theoretical physics with Max Born, with whom von Kármán collaborated in 1912 on lattice vibrations and their connection with specific heat. He demonstrated that the distance between vortices in the street, divided by the breadth of the street was equal to 1/0,238, in order to ensure stability.[22] This theoretical work, with H. Rubach's experiments [26], had important consequences and even today, the phenomenon known as alternating vortex shedding produced by a bluff body is often referred as the *von Kármán's street*.

Of course, Henri Bénard reacted to von Karman's supposed paternity on the alternating vortex emission by moving bodies in fluid even if they referred to one

Organized Bodies at the Collège de France from 1868 until his death. He is mainly known for his photographic experiments on the study of motion. In 1890, Marey continued to study aerodynamics by building a wind tunnel and taking pictures of coherent structures!

[22] T. von Kármán relates in his book [27] how he got interested in the subject, after L. Prandtl's suggestion.

of his 1908 papers [B12]. This is the reason why, for several years, Bénard argued indirectly, trying to demonstrate that von Kármán's and Rubach's theoretical results were idealistic and far from real conditions. An example can be found in the paper he published in 1926 entitled, *"About the Inaccuracy for Fluids in Real Conditions of the Kármán's Theoretical Laws Regarding Alternated Vortex Stability"* [B29]. Actually, Bénard experimentally showed that the dimensionless frequencies measured with the Strouhal number S, vary with the velocity of the object, whereas von Kármán's result could imply that Strouhal number is constant. As part of this polemic, Bénard insisted on the role played by the camera, an instrument that could reduce the fluctuations on period measurements and allow for better statistics. He quoted in a sarcastic way the 1912 von Kármán and Rubach paper: *"The authors said that it could be possible to use the cinematography,* [an obvious reference to his own work], *but such a method has no important advantage in respect with the method they used"*, and adding that the authors, giving results on only two experiments, counted vortices with a clock in hand, observing departure and arrival, *"method already used for horse racing, but which lacks precision for a 3 centimetres race course"* [B49]. In 1926, Bénard and von Kármán met at the Second International Congress of Applied Mechanics in Zurich. Bénard said[23] that von Karman kindly consented to him that the accuracy of the few experiments made by Rubach and himself in 1912 could not be compared to Bénard's ones. Moreover, Kármán declared that he never claimed paternity of the alternated vortex, with an exception, for the theory on its stability. As Bénard told the story, von Kármán declared pettishly that the expression *Kármánsche Wirbelstrasse* (Kármán Vortex Street) could be easily replaced by the *rue des tourbillons de Bénard* (Bénard Vortex Street). Kármán's version, in his book *The Wind and Beyond* [27], placed the same discussion in 1930 (the Third International Congress of Applied Mechanics held in Stockholm?) in the course of which he declared he would accept to call it *Kármán Street* in Berlin and London, and *rue de Bénard* in Paris.[24] Even if both said that the problem

---

[23] *"M. Th. von Kármán a bien voulu m'accorder très aimablement que les quelques expériences faites par Rubach et lui en 1912 ne pouvaient au point de précision être comparées aux miens. D'autre part, il a déclaré n'avoir jamais réclamé la priorité en ce qui concerne les tourbillons alternés, sauf pour sa théorie de leur stabilité. Il a ajouté avec humeur que l'expression die Kármánsche Wirbelstrasse pourrait sans incovenient être remplacée par la rue de tourbillons de Bénard. Mais, comme moi-même, M. Kármán trouve plus sérieux que la question de priorité, mon désaccord expérimental avec la loi de similitude dynamique, intangible aux yeux des hydrodynamiciens"* [B49].

[24] *I never asked to have my vortex theory named after me, but somehow the name remained. There is always some danger in such matters, especially as one grows in fame or importance. In 1930, almost two decades after my paper was published, a French professor named Henri Bénard popped up at an international congress and protested the name Kármán Vortex Street. He pointed out that he had observed the phenomenon earlier and that he had taken pictures of alternating vortices before I did. He was right, and as I did not wish to fight over names I said: "All right, I do not object if in London this is called Kármán Vortex Street. In Berlin let us*

of priority was not important, Bénard's insistence in spreading, even at the Paris University Council, the discussion which had taken place in Zurich in 1926, has to be underlined. His colleagues took care of his claim, as did H. Villat who, in a report to the Science Academy, criticized the fact that in a recent book Joukowski had dedicated a full chapter to von Karman, without any mention of Bénard's experiments. This attitude of Bénard about the conflict of paternity of the vortex street has to be placed in the general context of the First World War. We must keep in mind that in the immediate postwar period, scholars of the victorious powers developed an attitude of refusal towards German scholars and several academies of science decided to boycott the Germans from 1919 up to 1926 [29] . For example, French scientists refused to attend the conference on hydrodynamics and aerodynamics held in September 1922 in Innsbruck and organized by Levy and von Kármán. It is worthwhile to quote Marcel Brillouin (who had a great influence on Bénard's formation) who refused the invitation on August 8,1922, with a provocative letter: *"Since German scientists and professors have not understood that they would not be bound to pay if they had not committed systematic devastations, and because they have done it, they have to pay; my esteem for them is still insufficient to shake hands with, no matter their scientific value"*. Therefore, the strong insistence of Bénard to fight against von Kármán's priority (this is just a hypothesis) could have been fed by the strong chauvinistic climate of the time.[25] Even in 1924, the French were absent from the First International Congress of Applied Mechanics at Delft [30-32], organized by Burgers and Biezeno, under the impetus of von Kármán.[26] In addition, these years were characterized by two opposite styles of making science, especially in mathematics: the "modern" or formalist one represented by Gottingen University and Hilbert and the "conservative" or intuitive one by Poincaré and French science. Chauvinism of the time tried to transform this difference into a controversy.[27]

---

call it Karmansche Wirbelstrasse, and in Paris, Boulevard d'Henri Bénard". We all laughed heartily and Bénard and I became good friends"[27]. About the same event, von Karman said: ... *Bénard did a great deal of work on the problem before I did, but he chiefly observed the vortices in very viscous fluids or in colloidal solutions and considered them more from the point of view of experimental physics than from aerodynamics* [28].

[25] At that time, Paul Langevin was one of the few scientists in France who adopted a definite position against this boycott and tried to develop scientific cooperation with German scientists, in the first place, with A. Einstein [6].

[26] In 1913, von Kármán became director of the Aachen Aerodynamics Institute and in 1930, he moved to Caltech, in the USA. In 1937, he visited the Institute of Fluid Mechanics in Paris and after the Second World War played an important role in the French–American relationship. As an American delegate and Chairman of the AGARD (a scientific advisor group of NATO) he frequently visited Paris and developed a friendship with J. Pérès, L. Malavard, and other French scientists.

[27] But, irony of history, after 1933, these qualifications changed. In France, the Bourbaki group became the symbol of modern mathematics and, in Germany, formalism was attacked as a "Jewish science".

Fig. 2.10. H. Bénard (1) during the Third International Congress of Applied Mechanics , at Cambridge in 1934. Also attending this meeting were G.I. Taylor (2), L. Prandtl (3), and J. Burgers (4) *(IUTAM)*.

2.4 Bénard and Cinematography

Of course, Bénard was not the first one to use the cinematograph in science and we recall pioneers such as Étienne-Jules Marey, Georges Demenÿ, Lucien Bull, Louis and Auguste Lumière, and Georges Méliès [33]. But Bénard was one of the first to use it, a Lumière-Carpentier camera, as a scientific instrument, in 1907 in Lyon for his first experiments on vortex shedding. Before this, in 1900, he used a chronophotographic apparatus, supplied by L. Gaumont to take pictures each five seconds of the turbulent vortex. Bénard not only used cinematography as a technique of observation and measurement, but also as a means for scientific popularization, which perhaps is the least known aspect of his work. We have found in the archives of the Gaumont company, in the Joinville studios founded by Léon Gaumont in 1895 near Paris, a series of seven films[28] produced by Bénard and Dauzère between June and October 1913, entitled *"The Cellular Vortices"*, as part of another series called *"Cinematography Applied to the Study of Physical Science"*. On April 18, 1914, Dauzère and Bénard presented the films

---

[28] *Les tourbillons cellulaires: Particules solides et méthodes optiques* (H. Bénard, 8'28"); *Les tourbillons cellulaires du spermacéti: Leur régularisation progressive enregistrée par les méthodes optiques* (H. Bénard, 3'46"); *Les chaînes de tourbillons cellulaires de l'éther* (H. Bénard, 11'45"); *Les tourbillons cellulaires de l'éther: Relief de la surface libre* (H. Bénard, 5'49"); *Tourbillons cellulaires isolés: Observation par la méthode optique en lumière réfractée* (C. Dauzère, 4'40"); *Les deux espèces de tourbillons cellulaires* (C. Dauzère, 4'2"); *Solidification cellulaire* (C. Dauzère, 4'48").

in the amphitheatre of the Conservatoire des Arts et Métiers de Paris, to the French Physical Society. This series of films has been listed in the Gaumont catalogue since July 1914. We do not know if the films were actually shown in the schools, because the ones we have found were too long and lacked mounting and editing work, which leads us to think they were not considered, from the production point of view, as finished. The most important thing to emphasize is the interesting booklet [B19] added to the distribution of the films, probably meant for teachers, with a very synthetic description of convective motion phenomena intended for a wide public. In these films, the observation methods are explained and comments are developed on the analogies between convective motion and the biological phenomena, calling attention on the extraordinary beauty of descriptive physics. It is important to observe that the movies were produced by the Gaumont series on education. In fact, offering movies on craftwork, hygiene, zoology, and physics, Léon Gaumont considered himself a progressive educator, who thought that cinematography was an important educational support. In addition, Bénard participated in scientific popularization activities such as exhibitions. For instance, he presented a special convection container shaped to project a shadowgraph image on at exhibitions on physics [B42]. One of his assistants, M.R. Fabre told us that he built a wood model of a hexagonal cell representing convective motion that he showed in different activities and exhibitions.

Among the most spectacular contributions of Bénard and his team was the one at the Universal Exposition of Paris in 1937, called "Exposition Internationale des Arts et Techniques dans la vie moderne", which took place in the building later called the Palais de la Découverte, the first interactive science museum in the world. In the meteorology pavilion, Bénard's team presented experiments on convective cell production, where an air current was heating from below (what could be called the) and showing the longitudinal convective rolls, analogue to the clouds, in order to explain the morphology of different clouds: cirrocumulus, altocumulus and stratocumulus. It is worth mentioning that the presentation said, *"This organized pre-turbulent state, extremely important as a transition from the laminar regime to the turbulent chaotic regime, has been experimentally characterized at the stand on Meteorology, heating from below the air contained in a glass channel where the bottom is made from a mobile metallic tape and the other walls from glass"*. Visualizing the air trajectory with smoke, very beautiful cellular vortices were obtained with the static tape and, with the tape in movement, longitudinal rolls were produced. In the same stand, experiments were presented by Bénard's students, M. Luntz and D. Avsec who displayed electroconvective cellular vortices generated by a vertical gradient of electric potential. The exposition catalogue explained that the photos chosen by Bénard showed the most characteristic analogy between clouds in and , as well as the granulations of the solar photosphere with . Actually, during the 1930s, Bénard worked with the Commission of Atmospheric Turbulence, presided over by Philippe Wherlé, director of the National Meteorological Office, and a team of meteorologists and physicists. All together, they performed experiments on air with André Japy as pilot, showing the analogy between convective rolls and parallel cirrus bands, a topic which became the subject of one of Bénard's last works [B47, B48].

2.5 Physics and Dynamical Systems

Bénard was a council member of the French Physical Society and president from 1928 to 1929, following Louis Lumière and preceding Jean Perrin, Nobel Prize in Physics in 1926. In his investiture address, Perrin recognized Bénard's role, *"He has given to hydrodynamics the methods of the Physics, doing so, he discovered and studied phenomena not foreseen by theoreticians"*. Bénard himself confessed in 1929, in his inaugural address in the Institute of Fluid Mechanics of Paris [B41], that after his earlier work, he became interested in hydrodynamics and adding that *"he was like Monsieur Jourdain, who was writing prose without knowing it."* In the same speech, he said that he had studied the classical treatises of Lamb and Basset, comparing them with Lord Kelvin's and Lord Rayleigh's, which he considered to have been written by physicists who had maintained their link with reality. From his point of view, Lamb and Basset's treatises were concerned with pretty much theoretical and mathematical physics and did not deal enough with experiments: there were no pictures, so one was led to the wrong conclusion that physics was nothing but equations. He therefore had the idea of making a photo album on fluid motion with images taken from photographs and films in order to illustrate theoretical works. (This was actually done many years later in Van Dyke's famous Album of Fluid Motion).

Because Bénard's works inspired, as shown in these proceedings, most of the modern experimental works on dynamical systems and , one should quote a very interesting work performed by Bénard's group, in this direction. François-Joseph Bourrières (1880-1970), who was a former undergraduate student of Bénard and Duhem in Bordeaux,[29] collaborated in Bénard's laboratory in the Institute of Fluid Mechanics, on the subject of fluid–structure interaction. Bénard explained that he trained him in *"the experimental and rational analysis of the phenomena of and oscillation existing in real fluid mechanics"* a topic on which he had already drawn attention after 1900. Indeed, one of Bourrières' subjects was the study of the movements of the free end of a flexible rubber tube, inside which a fluid was circulating. Bourrières described spontaneous oscillations as well as

---

[29] The role of fluctuations in dynamical systems was Duhem's concern. In 1898, W.S. Franklin wrote in the Physical Review [34] a review of Duhem's book, titled *Traité élémentaire de mécanique chimique* published a year before in Paris. Impressed by Duhem's analysis, he said ," *A state of unstable equilibrium is produced by the heating of the lower strata of the atmosphere. An infinitesimal action as the waving of a fan, may precipitate a sweep,.... in other words, an infinitesimal cause may produce a finite effect. Long range detailed weather prediction is therefore impossible... **the accuracy of this prediction is subject to the condition that the flight of a grasshopper in Montana may turn a storm aside from Philadelphia to New York.***" This reference is cited by R.C. Hilborn in a very recent study of the evolution of the notion of sensitive dependence on initial conditions and chaos [35].

more complex ones. In order to study the phenomenon through experiments, he attached a little lamp at the extremity of the tube and took pictures, with an open shutter, of their trajectory. He then could display a very clear picture of a limit cycle and the attraction to this cycle limit during transitional phase and perturbations. Bourrières published this experiment in 1939, with a foreword by Bénard [36], in which the latter exposed the nonperiodic form the movement could take, from whatever initial conditions, and how these conditions led the system to a limit cycle, referring to the concept of self-oscillations due to Alexandr A. Andronov (1901-1952) [37]. As it is known, one of great achievements of Andronov, who belonged to the group on nonlinear dynamics formed in Moscow around Leonid I. Mandelstam (1879-1944), was to demonstrate, in the late 1920s, the connection between Poincaré's limit cycles and a whole range of practical oscillatory processes.

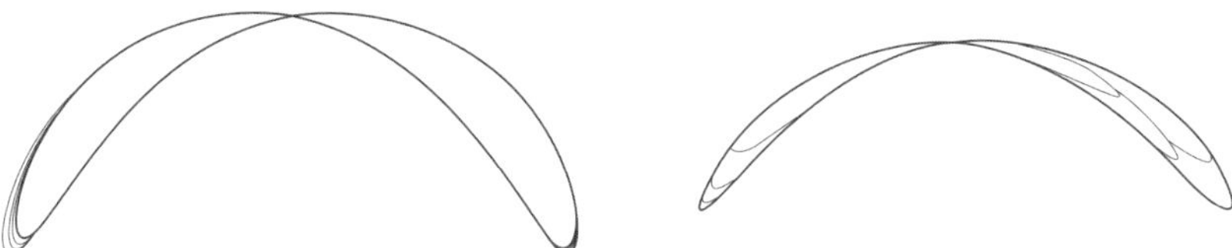

Fig. 2.11. Physical limit cycle and transients, in Bourrières' experiment [36].

2.6 Conclusion

This scientific biography provided us with the opportunity to bring to light the remarkable homogeneity of Henri Bénard's works in experimental hydrodynamics, especially on the study and recording of the movement of liquids through optical means. On two occasions he discovered, in this field, new phenomena or phenomena hardly suspected before him. He found the geometric or kinematics laws of convective cells in 1899 and of the alternating vortex shedding behind bluff bodies in motion, in 1906 to 1907. These studies, realized at the very beginning of his career, were actually his most important contributions.

This scientific biography, allows us to guess how, in France after the Second World War, experimental fluid dynamics, topics belonging to physics, withdrew from it but, in the late seventies, came back, exactly around the topics Bénard had previously explored.

Far from being a review of Bénard's experiments, this biography nevertheless permits us to understand the context his work was achieved in as well as the limits its developments had suffered. Some science historians are nowadays wondering at the gap between founding works such as Bénard's and modern studies on physical hydrodynamics in dissipative structures. We hope this chapter will contributes to the reduction of this very gap.

Acknowledgments

I would like to especially acknowledge the collaboration of Robert Fabre, who was Bénard's technical assistant from 1930 until 1939. I benefited from the help and collaboration of M. Monnerie (ESPCI), M. Morelle (Rectorat, Paris), M. Hollier (FAST-Orsay), T. Shinn (MSH-CNRS), Y. Roccard (Paris), F. Tully (Observatoire Nice), G. Battimelli (Universita di Roma), C. Fontanon (CNAM), L. Rosenthal (Paris), E. de Langre (Ladhyx), E. Guyon (ESPCI), F. Auger (ESPCI), J.F. Stoffel (Liège), and also by the Palais de la Découverte and the Centre de Ressources Historiques of ESPCI.

A. General References

1. A. Foch, *Annuaire de la Fondation Thiers*, 23–36 (1939/1940).
2. C. Charles and E. Telkes, *Les professeurs de la Faculté des sciences de Paris: Dictionnaire biographique 1901-1939*, Ed. INEP-Editions du CNRS (1989).
3. P.L. Schereschewsky, Le soixante-quinzième anniversaire des cellules atmosphériques de Bénard (1901-1976), *J. de Recherches Atmosphériques* Vol. X, 1–7 (1976).
4. *Le centenaire de l'Ecole Normale 1795-1895*, Ed. Presses de l'ENS, Paris (1994).
5. C. Zwerling, The Emergence of the Ecole Normale Superieure as a center of scientific education in the nineteenth century in: *The Organization of Science and Technology in France 1908-1914*, R Fox & G Weisz,Eds.: Cambridge University Press, Cambridge, UK (1980).
6. J. Langevin, *Paul Langevin, mon père*, Ed. Français Réunis, Paris (1971).
7. M. Pietrocola Pinto de Oliveira, *E. Mascart et l'optique des corps en mouvement*. Ph.D. thesis, Université Paris VII (1992).
8. R. Mosseri, *Léon Brillouin: A la croisée des ondes*, Ed. Belin, Paris (1998).
9. M. Brillouin, Expériences de Bénard, *Leçons sur la viscosité des liquides et des gaz*, Vol.1, 152–159, Paris (1907).
10. S.L. Jaki, *Reluctant heroine: The life and work of Hélène Duhem*, Ed. Scottish Academic Press, Edinburgh (1992).
11. C. Fontanon, *Histoire de la mécanique appliquée: Enseignement, recherche et pratiques mécaniciennes en France après 1880*, Ed ENS Editions, Lyon 1998 and *Painlevé, homme d'Etat et promoteur des sciences de l'aviation (1910-1933)*, to be published.
12. P.E. Mounier-Kuhn, L'enseignement supérieur, la recherche mathématique et la construction de calculateurs en France (1920-1970) in *Des ingénieurs pour la Lorraine*, F. Birck and A. Grelon, Eds. Serpenaise, Cambridge (1998).
13. M. Roy, Notice nécrologique de H. Villat, *C. R. Acad. Sci.*, **274**, 127–132, 24 avril 1972.
14. P. Germain, Joseph Pérès et le renouveau de la mécanique en France, *C.R. Acad. Sci.*, 12 décembre 1977.

15. D. Pestre, *Physique et physiciens en France* 1918-1940, Ed. EAC, Paris (1992).
16. P. Idrac, Sur les courants de convection dans l'atmosphère dans leur rapport avec le vol à voile et certaines types de nuages, *C.R. Acad. Sci.*,**171**, 42–44, 5 juillet 1920.
17. E. Davoust, *L'Observatoire du Pic du Midi (Cent ans de vie et de science en haute montagne)* CNRS-Editions, Paris (2000).
18. E.L. Koschmieder, *Bénard Cells and Taylor Vortices*, Ed. Cambridge University Press, Cambridge, UK (1993).
19. M.J. Block, Surface tension as the cause of Bénard cells and surface deformation in a liquid film, *Nature* **178**, 650–651 (1956).
20. J.R.A. Pearson, On convection cells induced by surface tension, *J. Fluid Mech.* **4**, 489–500 (1958).
21. D.A. Nield, Surface tension and bouyancy effects in cellular convection, *J. Fluid Mech.* **19**, 341–352 (1964).
22. C. Dauzère, Sur la stabilité des tourbillons cellulaires, *C.R. Acad. Sci.*, **154**, 974–977, 15 avril 1912.
23. Lord Rayleigh, Aeolian tones, *Phil. Mag.* Ser. 6 **29**, 433-444 (1915).
24. H.R.A. Mallock On the resistance of air, *Proc. Roy. Soc.*, 262–273, 7 March 1907.
25. E. Marey, *Les mouvements de l'air étudiés par la chronophotographie*, La Nature 232-234 (1901) and P. Noguès, *Recherches expérimentales de Marey sur le mouvement dans l'air et dans l'eau*, Publications scientifiques et techniques du Ministère de l'Air **25** (1933).
26. T. von Kármán, and H. Rubach, Über den Mechanismus des Flüssigkeits und Luftwiderstandes *Physikalische Zeitschrift*,**13**, Heft2; 49–59 (1912). English version: On the resistance mechanisms of fluids and air *NASA-TT-F-16415* (1975).
27. T. von Kármán, with L. Edson, *The Wind and Beyond* Ed. Little, Brown, Boston (1967).
28. T. von Kármán, *Aerodynamics*, Ed. Cornell University Press, Ithaca (1957).
29. B. Schroeder-Gudehus, *Les scientifiques et la paix: La communauté scientifique internationale au cours des années 20*, Ed. Presses de l'Université de Montréal, Montréal (1978).
30. S. Juhansz, *IUTAM - A short history*, Ed. Springer Verlag, Berlin (1988).
31. W. Schiehlen and L. van Wijngaarden, *Mechanics at the turn of the century*, Ed. Shaker Verlag, Aachen (2000).
32. G. Batchelor, *The life and legacy of G.I. Taylor*, Ed. Cambridge University Press, Cambridge, UK (1996).
33. N. Brenez and C. Lebrat, *Jeune, dure et pure! Une histoire du cinéma d'avant-garde et expérimental en France*, Ed. Cinémathèque Française/ Mazzotta, Paris-Milan (2000).
34. W.S. Franklin, "Book Review: Traité Elémentaire de Mécanique Chimique fondée sur la Thermodynamique", *Physical Review* **VI**, 170–175 (1898).

35. R.C. Hilborn, Sea gulls, butterflies, and grasshoppers: A brief history of the butterfly effect in nonlinear dynamics, *Am. J. Phys.* **72**, 425–427 (2004).

36. F.J. Bourrières, Sur un phénomene d'oscillation auto-entretenue en mécanique des fluides réels, *Publications Scientifiques et Techniques du Ministère de l'Air*, N° **147** (1939).

37. A.A. Andronov, Les cycles limites de Poincaré et la théorie des oscillations auto-entretenues, *C.R. Acad. Sci.*, **189**, 559–561(1929).

B. Bénard Papers

1. H. Bénard et E. Mascart, Sur le pouvoir rotatoire du sucre, *Ann. Ch. et de Phys.*, 7è série, **17**, 125–144 (1899).

2. H. Bénard et L.J. Simon, Sur les phénylhydrazones du d-glucose et leur multirotation, *C.R. Acad. Sci.*, 4 mars 1901.

3. H. Bénard, Etude expérimentale du mouvement des liquides propageant de la chaleur par convection. Régime permanent: Tourbillons cellulaires, *C.R. Acad. Sci.*, **130**, 1004–1007, 9 avril 1900.

4. H. Bénard, Mouvements tourbillonnaires à structure cellulaire. Etude optique de la surface libre, *C.R. Acad. Sci.*, **130**, 1065–1068, 17 avril 1900.

5. H. Bénard, Tourbillons cellulaires dans une nappe liquide, $1^{\text{ère}}$ partie. Description générale des phénomènes, *Revue Générale des Sciences pures et appliquées*, **11**, 1261–1271, 15 décembre 1900.

6. H. Bénard,Tourbillons cellulaires, etc., $2^{\text{ème}}$ partie. Procédés mécaniques et optiques d'examen; lois numériques des phénomènes, *Revue Générale des Sciences pures et appliquées*, **11**, 1309–1328, 30 décembre 1900.

7. H. Bénard,Etude expérimentale des courants de convection dans une nappe liquide. Régime permanent: tourbillons cellulaires, *C.R. de l'Ass. pour l'Avanc. des Sciences*, 446–467, Congrès de Paris (1900).

8. H. Bénard, Les tourbillons cellulaires dans une nappe liquide propageant de la chaleur par convection en régime permanent, *Ann. de Ch. et de Phys.* (1901).

9. H. Bénard, Les tourbillons cellulaires dans une nappe liquide propageant de la chaleur par convection en régime permanent. *Thèse de doctorat*, Gauthier-Villars, 1–88 (1901).

10. H. Bénard, Etude expérimentale des courants de convection dans une nappe liquide. Régime permanent: Tourbillons cellulaires, *Journ. de Phys.*, $3^{\text{ème}}$ série, **9**, 513 et **10**, 254 et suivantes, 1900 et 1901.

11. H. Bénard, Etude expérimentale des courants de convection dans une nappe liquide. Régime permanent: tourbillons cellulaires, *Bull. des séances de la Société Française de Physique*, 1901.

12. H. Bénard, Formation périodique de centres de giration à l'arrière d'un obstacle en mouvement, *C.R. Acad. Sci.*, **147**, 839–842, 9 novembre 1908.

13. H. Bénard, Etude cinématographique des remous et des rides produits par la translation d'un obstacle, *C.R. Acad. Sci.*, **147**, 970–972, 23 novembre 1908.

14. H. Bénard, Sur les tourbillons cellulaires, *Ann. de Ch. et de Phys.*, 8ème série, **24**, 563–566, décembre 1911.

15. H. Bénard, Sur la formation des cirques lunaires, d'après les expériences de C. Dauzère, *C.R. Acad. Sci.*, **154**, 260–263, 29 janvier 1912.

16. H. Bénard, Sur le clivage prismatique dû aux tourbillons cellulaires (amidon, basalts, etc), *C.R. Acad. Sci*, **156**, 882–884, 17 mars 1913.

17. H. Bénard, Sur la zone de formation des tourbillons alternés derrière un obstacle, *C.R. Acad. Sci.*, **156**, 1003–1005, 31 mars 1913.

18. H. Bénard, Sur la marche des tourbillons alternés derrière un obstacle, *C.R. Acad. Sci.*, **156**, 1225–1228, 21 avril 1913.

19. H. Bénard et M.C. Dauzère, *Les tourbillons cellulaires, notice sur une série de films obtenus dans les laboratoires Gaumont,* juillet-octobre 1913, Société des Etablissements Gaumont, éditeur, 1914.

20. Quatre films (de H. Bénard) décrits dans la notice précédente : Tourbillons cellulaires du spermaceti (2 films); tourbillons cellulaires (en chaînes) de l'éther s'évaporant (2 films).

21. H. Bénard, Détermination du coefficient de transmission de la chaleur à travers les parois d'un wagon fortement isolé, Soc. Fr. de Phys. Résumé des communications faites en 1919, séance du 17 janvier 1919.

22. H. Bénard, Sur l'utilité de la lumière polarisée dans les observations faites en mer ou au bord de la mer, et sur une jumelle à polariseurs. Bull. off. de la Dir. des Recherches scient. et ind. et des Inventions, **4**, 229–248, février 1920.

23. H. Bénard, Expériences et mesures techniques relatives aux wagons aménagés en France pour le transport de la viande congelée aux armées. *Revue générale du Froid et des Industries frigorifiques*, **2**, numéro de mars, avril, mai 1921, Réimpression, I brochure 24 pages in-4°.

24. H. Bénard, Note sur une lunette panoramique à verres cylindriques, anastigmatique, monoculaire et binoculaire, *Bull. des Inventions*, **21**, 426–429, juillet 1921.

25. H. Bénard, Dispositifs optiques pour montrer en projection les petits défauts d'une surface presque plane ou les petites hétérogénéités d'indice de réfraction d'une lame à faces parallèles, solide ou liquide, etc. Applications : Tourbillons cellulaires, etc., *J. de Phys. Radium*, **4**, 495–496, décembre 1923.

26. H. Bénard, Sur les tourbillons alternés derrière un obstacle, *J. de Phys. Radium*, **4**, 497 (1923).

27. H. Bénard, Dispositifs optiques pour projeter sur un écran les tourbillons cellulaires, et plus généralement les petites hétérogénéités d'indice de réfraction et d'épaisseur que peut présenter une nappe liquide, etc. *Revue d'optique théorique et expérimentale*, **3**, 127–136, mars 1924.

28. H. Bénard, Sur les lois de la fréquence des tourbillons alternés détachés derrière un obstacle, *C.R. Acad. Sci.*,**182**, 1375–1377, 7 juin 1926.

29. H. Bénard, Sur l'inexactitude, pour les liquides réels, des lois théoriques de Karman relatives à la stabilité des tourbillons alternés, *C.R. Acad. Sci.*,**182**, 1523–1525, 21 juin 1926.

30. H. Bénard, Sur les écarts des valeurs de la fréquence des tourbillons alternés par rapport à la loi de similitude dynamique, *C.R. Acad. Sci.*,**183**, 20–22, 5 juillet 1926.

31. H. Bénard, Sur la limite du régime laminaire et du régime turbulent, révélée par l'apparition de tourbillons alternés nets, *C.R. Acad. Sci.*,**183**, 184–186, 19 juillet 1926.

32. H. Bénard, *Notes sur les tourbillons alternés derrière un obstacle*, Réimpression des huit notes précédentes, avec une introduction et des additions, 1926.

33. H. Bénard, *Notes sur les tourbillons alternés derrière un obstacle*, 23 et 24. Communications présentées au deuxième Congrès International de Mécanique appliquée, Section d'Hydrodynamique, Zürich, septembre 1926, Ed. Orell Füssli 1927.

34. H. Bénard, *Sur les tourbillons alternés détachés derrière un obstacle en mouvement,* Bull Soc. Franc. de Physique **238**, 118–121, 3 décembre 1926.

35. H. Bénard, Sur les tourbillons cellulaires et la théorie de Rayleigh, *C.R. Acad. Sci.*, **185**, 1109–1111, 21 novembre 1927.

36. H. Bénard, Sur les tourbillons en bandes et la théorie de Rayleigh, *C.R. Acad. Sci.*, **185**, 1257–1259, 5 décembre 1927.

37. H. Bénard, Sur les tourbillons alternés dus à des obstacles en lames de couteau, *C.R. Acad. Sci.*, **187**, 1028–1030, 3 décembre 1928 (and erratum in *C.R. Acad. Sci.*, **187**, 1300).

38. H. Bénard, Les tourbillons alternés et la loi de similitude dynamique, *C.R. Acad. Sci.*, **187**, 1123–1125, 10 décembre 1928.

39. H. Bénard, Les tourbillons des bains développateurs, *Bull. Soc. Franc. de Physique* **266**, 1928, 112S.

40. H. Bénard, Tourbillons alternés à la surface d'une nappe liquide fendue par un obstacle, *Bull. Soc. Franc. de Physique* **271** (1929).

41. H. Bénard, Mécanique expérimentale des Fluides, *Revue Scientifique* **11S**, 28 décembre 1929.

42. H. Bénard, Appareil destiné à l'étude des sillages hydrodynamiques par la méthode des rides, cinématographiant à grande fréquence un champ circulaire de 24 cm. de diamètre, *Proceedings of the Third International Congres of Applied Mechanics*, Stockholm 1930; also Rév. Aéronautique (1930).

43. H. Bénard, Anamorphoseurs, *Revue d'optique théorique et expérimentale*, **9** (1930).

44. H. Bénard, Sur la limite théorique de stabilité de l'équilibre préconvectif de Rayleigh, comparée à celle que donne l'étude expérimentale des tourbillons cellulaires de Bénard *Proceedings of the Fourth International Congres of Applied Mechanics*, Cambridge (1934) and Journées de Mécanique des Fluides de l'Université de Lille t. II, 317 Ed. Chiron (1935).

45. H. Bénard, La photosphère solaire est-elle une couche de tourbillons cellulaires? *C.R. Acad. Sci.* **201**,1935, 1328–1330 and also in *Volume jubilaire offert à Marcel Brillouin* Ed. Gauthier-Villars, Paris, 124–137 (1935).

46. H. Bénard, Revue des travaux expérimentaux récents sur la formation des nuages en bandes et des nuages en balles de coton (Mammatocomulus) en

liaison avec la théorie des tourbillons dus à la convection de la chaleur (tour-billons cellulaires polygonaux et tourbillons en bandes), *Proceedings General Assembly Int. Union of Geodesy and Geophys*, Edinbourg (1936).

47. H. Bénard, P. Idrac et la théorie des nuages en bandes, *La Météorologie* **3**,1–4 (1936).

48. H. Bénard et D. Avsec, Travaux récents sur les tourbillons cellulaires et les tourbillons en bandes: Applications à l'astrophysique et à la météorologie, *J. de Phys. Radium* **9**, 486–500 (1938).

49. *Notice sur les Titres et Travaux Scientifiques de M. Henri Bénard*, Ed. Gauthier-Villars, Paris 1926 and 1929.

Part II

Bénard Cellular Structures

3 Rayleigh–Bénard Convection: Thirty Years of Experimental, Theoretical, and Modeling Work

Paul Manneville

Laboratoire d'Hydrodynamique, École Polytechnique
F-91128 Palaiseau cedex, France
`paul.manneville@ladhyx.polytechnique.fr`

A brief review of Rayleigh–Bénard studies performed during the twentieth century is presented, with an emphasis on the transition to turbulence and the appropriate theoretical framework, relying on the strength of confinement effects and the distance to threshold, either dynamical systems for temporal chaos in the strongly confined case, or models of space–time chaos when confinement effects are weak.

3.1 Introduction

The idea of convection is quite old (Hadley, Lomonossov, Rumford, etc.) but the first quantitative experiments were performed by Henri Bénard around year 1900 [2]. Figure 3.1 is a low-resolution reproduction of one of his original photographs. In fact Bénard studied the stability of a thin fluid layer open to air and submitted to a vertical temperature gradient. He accurately determined properties such as the space periodicity of the hexagonal pattern, its variation, and the profile of the interface. Later, in 1916, Lord Rayleigh [4] proposed his theory of a feedback coupling resting on buoyancy: a fluid particle hotter than its environment encounters ever colder fluid as it rises, which leads to the instability, as sketched in Figure 3.2 (left). He developed a complete linear stability analysis assuming stress-free conditions for the velocity and good heat-conducting plates. This mechanism was accepted as the explanation of Bénard's results until the role of the thermal Marangoni effect was pointed out, in particular by Pearson [3]: a temperature fluctuation at the surface induces tangential stresses that can be amplified by hot fluid coming from the inside, as suggested in Figure 3.2 (right). Surface tension usually increases as the temperature decreases while surface elements with the larger want to shrink, so that the mechanism indeed works when the temperature gradient is directed towards the interior of the layer. Some of the confusion may be explained by the fact that hexagons are usually expected when the top to bottom symmetry is broken, that is, when boundary conditions at the top plate are different from those at the bottom plate, which was the case of Bénard's experiments with a free upper surface. In that case, the two mechanisms are in competition but a bump is implied at the place where the fluid rises when the buoyancy is involved, whereas the effect of surface tension

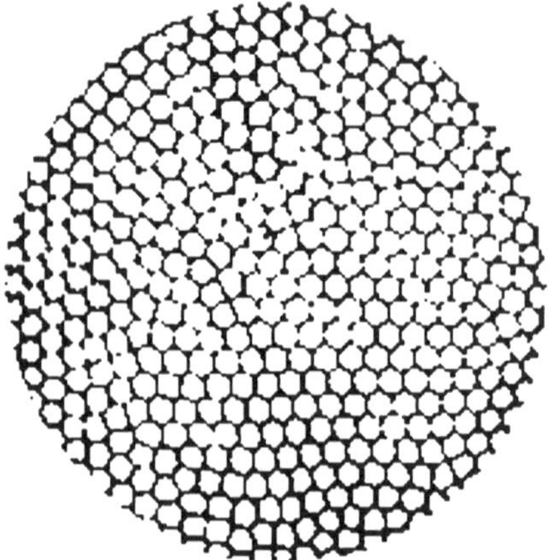

Fig. 3.1. One of Bénard's celebrated original photographs of the top view of convection patterns in a thin layer of spermaceti heated from below.

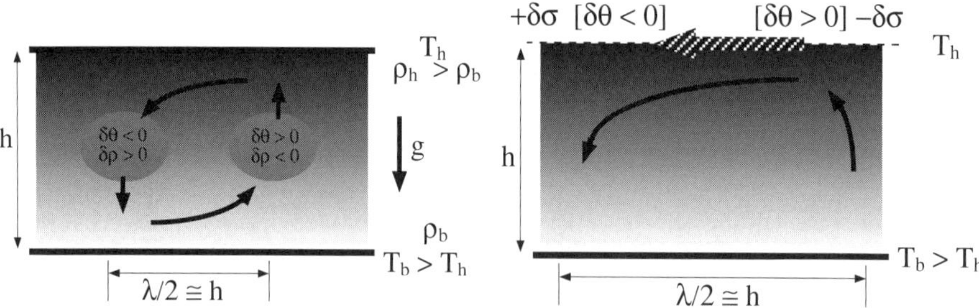

Fig. 3.2. Mechanisms for Bénard convection: (left) Motion of fluid is self-sustained as soon as gravitational energy release can overcomes dissipation losses (Rayleigh's idea); (right) The surface-tension mechanism (Marangoni effect) works even in zero-gravity environment provided that hot fluid comes from inside.

is to create a dip. Buoyancy effect in the bulk will dominates in thick layers and surface tension in thin ones so that there exists a compensation thickness at which both deformations balance each other [4]. This being recognized, in the following I restrict myself to the consideration of convection between solid plates according to Rayleigh's mechanism, hereafter called *RB convection*, as opposed to *BM*, where "R" stands for Rayleigh, "B" for Bénard, and "M" for Marangoni.

I briefly review the results of linear stability analysis in Section 3.2, and then some early nonlinear findings about secondary instabilities and the transition to turbulence in Section 3.3. After these preliminaries I mainly turn to modeling issues raised by the theoretical understanding of the results beyond the threshold. As to weakly turbulent states developing moderately far from the threshold, I insist on the role of confinement effects in controlling the nature of scenarios either towards chaos as envisioned in dynamical systems theory (Section 3.4) or towards space–time chaos (Sections 3.5 to 3.7). Before closing the Chapter, I also say few words in Section 3.8 about the account of strongly turbulent sates when the applied temperature gradient becomes arbitrarily large.

The present review is of course sketchy and biased by my personal interests. For further information and complementary views, the reader is invited to consult the works mentioned in the bibliography. An early self-contained presentation can be found in the first chapters of Chandrasekhar's book [1]. A more recent general reference is by Koschmieder [16]. Consult also the reviews by Busse [7] about nonlinear convection, and by Newell, Passot, and Lega [8] and Cross and Hohenberg [9] for the theoretical approach of space–time chaos with emphasis on the envelope formalism and general aspects of space–time chaos, respectively. The most recent developments regarding the transition to turbulence in convection can be found in the review by Bodenschatz, Pesch, and Ahlers [24], and the article by Siggia [11] is devoted to fully developed turbulent convection.

3.2 RB Convection at Threshold

Rayleigh's theory for the instability threshold [1] was developed within the so-called *Boussinesq approximation*, that is, the Navier–Stokes equations for an incompressible flow completed by the energy equation generalizing the Fourier diffusion equation to a fluid medium. The simplifying assumption is that the temperature only enters the state equation to account for thermal expansion, and all other fluid parameters are kept constant. Within this approximation the governing equations read:

$$\boldsymbol{\nabla}_h \cdot \mathbf{v}_h + \partial_z v_z = 0 \tag{3.1}$$

$$P^{-1}\left[\left(\partial_t + \mathbf{v}\cdot\boldsymbol{\nabla}\right)\mathbf{v}_h + \boldsymbol{\nabla}_h p\right] = \boldsymbol{\nabla}^2 \mathbf{v}_h \tag{3.2}$$

$$P^{-1}\left[\left(\partial_t + \mathbf{v}\cdot\boldsymbol{\nabla}\right)v_z + \partial_z p\right] = \left(\boldsymbol{\nabla}^2 v_z + \theta\right) \tag{3.3}$$

$$\left(\partial_t + \mathbf{v}\cdot\boldsymbol{\nabla}\right)\theta = \boldsymbol{\nabla}^2 \theta + R\,v_z \tag{3.4}$$

These equations are written here in dimensionless form for the perturbation around the basic state. Coordinate z is along the vertical and the subscript h indicates the horizontal direction. The temperature fluctuation θ is defined as the departure from the linear temperature profile given by the Fourier law in the fluid at rest: $T_0(z) = T_b - z\Delta T/h$ (h is the height of the cell and $\Delta T = T_b - T_t$ is the temperature difference between the bottom and top plates). The natural control parameter is the *Rayleigh number* defined as:

$$R = \frac{\alpha \Delta T g h^3}{\kappa\nu},$$

where α is the expansion coefficient and g the acceleration of gravity. The thermal diffusivity κ and the kinematic viscosity ν parameterize the stabilizing dissipative

---

[1] The material in this section is discussed at length in Chandrasekhar's book [1].

processes. The last parameter that shows up in these dimensionless equations is the Prandtl number $P = \nu/\kappa$ that controls the nature, either mostly thermal or rather hydrodynamic, of the physical processes at stake, as they depend on the relaxation rates of temperature and vorticity fluctuations. Equation (3.1) accounts for the continuity of the flow and (3.4) for heat conduction in the fluid. Differential buoyancy shows up in (3.3) through the term in θ and, at this stage the horizontal component of the Navier–Stokes equation (3.2) only plays a passive role (closing of flow lines). Boundary conditions must be added to these equations. The cleanest situation is for good conducting rigid (i.e., nonslip) plates, which yields:

$$\theta(z = z_\mathrm{p}) = 0 \quad \text{and} \quad \mathbf{v}(z = z_\mathrm{p}) = \mathbf{0}, \tag{3.5}$$

where z_p denotes the plate's position. As already mentioned, Rayleigh assumed stress-free boundary conditions at top and bottom, hence:

$$\theta(z = z_\mathrm{p}) = 0 \quad and \quad v_z(z = z_\mathrm{p}) = 0, \quad \boldsymbol{\nabla}_\mathrm{h}\mathbf{v}_\mathrm{h}(z = z_\mathrm{p}) = \mathbf{0}. \tag{3.6}$$

He solved the problem for Fourier normal modes in the form $\propto \exp(i\mathbf{k}_\mathrm{h} \cdot \mathbf{x}_\mathrm{h})$ where all the functional dependencies in the vertical coordinate are absorbed in the proportionality sign. A straightforward calculation from the Boussinesq equations (3.1 to 3.4) and boundary condition (3.6) yields the marginal stability condition:

$$R_\mathrm{m}(k) = \frac{\left(k^2 + \pi^2\right)^3}{k^2},$$

where k is the length of the horizontal wavevector $\mathbf{k}_\mathrm{h}$. This curve is displayed in Figure 3.3, together with that corresponding to the more realistic no-slip velocity boundary conditions (3.5) obtained by Pellew and Southwell (1940). The minimum of each curve defines the corresponding threshold above which convection sets in, $R_\mathrm{c} = 27\pi^4/4 \simeq 657.5$ and $R_\mathrm{c} \simeq 1708$ for stress-free and no-slip conditions, respectively. The expected diameter of the convection cells is half the critical wavelength $\lambda_\mathrm{c} = 2\pi/k_\mathrm{c}$. In the no-slip case $k_\mathrm{c} \simeq 3.12$, which makes the predicted diameter very close to the height of the cell ($\pi/3.12 \simeq 1.007$). In the stress-free case one obtains $k_\mathrm{c} = \pi/\sqrt{2}$ which gives a somewhat larger diameter ($\sqrt{2} \simeq 1.4$). This prediction could have partly been responsible for the confusion alluded to above because it was close to Bénard's observations (but for a Marangoni-based mechanism) although, strictly speaking, one should have compared it with the prediction for mixed boundary conditions (bottom: no-slip, top: free) that yields $R_\mathrm{c} \simeq 1101$ and $k_\mathrm{c} \simeq 2.68$. (Of course, other cases can also be studied, especially by relaxing the assumption about the thermal conductivity of the plates.)

3.3 RB Convection Beyond Threshold

As the Rayleigh number exceeds the threshold value, part of the heat is transported by convection which decreases the potential for instability, and thermal

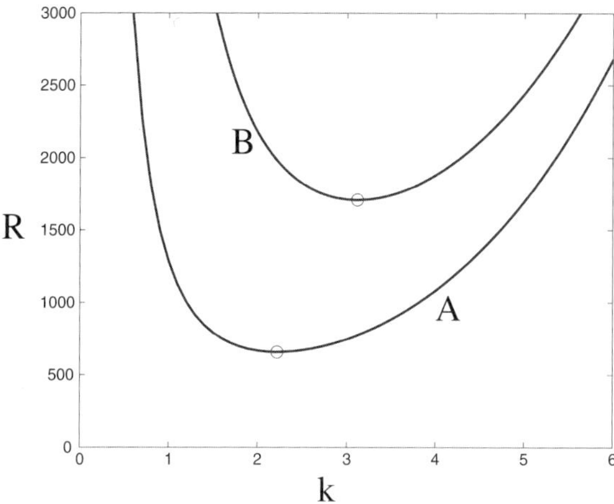

Fig. 3.3. Marginal stability curves for stress-free (A) and no-slip (B) velocity boundary conditions and isothermal plates (infinitely large conductivity).

diffusion and viscous friction increase due to the horizontal gradients implied by the modulation. This explains that the amplitude of the motions saturates beyond threshold. For rolls, the bifurcation turns out to be *supercritical*, that is, behaves continuously in the vicinity of the threshold and the system builds up a well-defined steady pattern. Early studies have been devoted to the problem of which kind of periodic pattern was selected by nonlinearities, rolls made of a single pair of wavevectors $\pm k\hat{\mathbf{x}}$ where $\hat{\mathbf{x}}$ is a unit vector in the horizontal plane, squares made of two pairs of wavevectors at right angles, and hexagons with three pairs at 120° (Figure 3.4). The theory was again first developed for stress-free boundary conditions (Malkus and Veronis, 1958 [12]) and later in the no-slip case (Schlüter, Lortz, and Busse, 1965 [38]), both studies concluding for roll patterns. Subsequent developments concerning the stability of the rolls against various secondary mechanisms mainly belong to Busse and Clever [7]. The result is a surface in parameter space called the *Busse balloon* separating stable roll patterns from unstable ones. In addition to the Rayleigh number, stability depends on the wavelength of the cells and on the Prandtl number. Some secondary modes are universal; that is, they do not depend on the fact that the pattern is generated by the RB mechanism but on the symmetries of the rolls (invariance through translation → Eckhaus instability, and rotation → zigzag instability) or on the fact that the intensity of the convection is weak close to the marginal curve (cross-roll instability). Other secondary modes are much more specific to convection, with structures that strongly depend on the value of P, for example the bimodal instability with secondary rolls localized in thermal boundary layers at right angles with the primary pattern when P is large. Figure 3.5 displays a picture of the Busse balloon.

Secondary instabilities are just a step towards more complex behavior as R is increased. Different scenarios have indeed been observed, depending on the

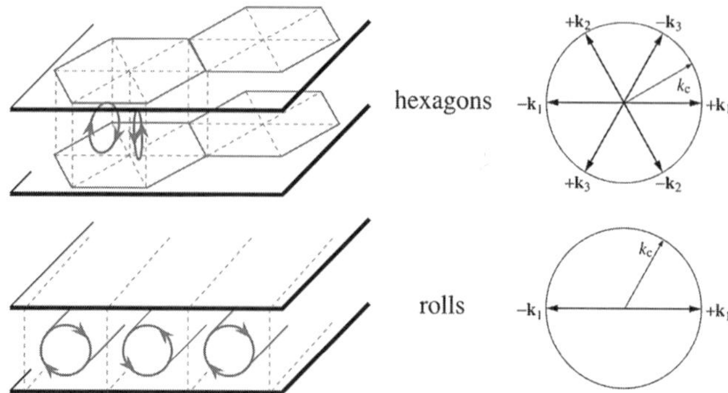

Fig. 3.4. Hexagons *versus* rolls. Hexagons result from the superposition of three pairs of modes at 120°. Those appearing in the thin convecting layer are produced by the Marangoni effect and have fluid sinking in the center of the cells. In thicker layers, the Rayleigh mechanism produces hexagons with fluid sinking at the cell edges. Both bifurcate subcritically. Rolls are obtained from a pair of modes with opposite wavevectors that may point to any direction in the horizontal plane owing to orientational degeneracy in laterally unbounded layers. This degeneracy is then broken by the developing mode. Rolls bifurcate supercritically.

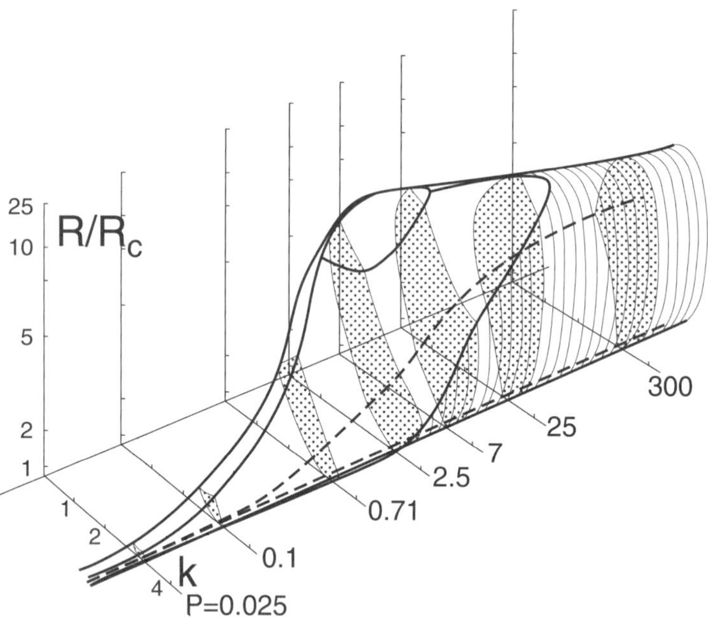

Fig. 3.5. Busse balloon in perspective in the (k, P, R)-space (after Busse [7]).

value of P. Beyond the primary instability leading to the formation of time-independent two-dimensional[2] rolls, time dependence was observed to introduce itself first and at relatively low R when P is small, but only after secondary instabilities adding space dependence along the rolls (three-dimensional time independent states) at large P at higher R. A compilation of early experimental results, adapted from Krishnamurti [28], is displayed in Figure 3.6.

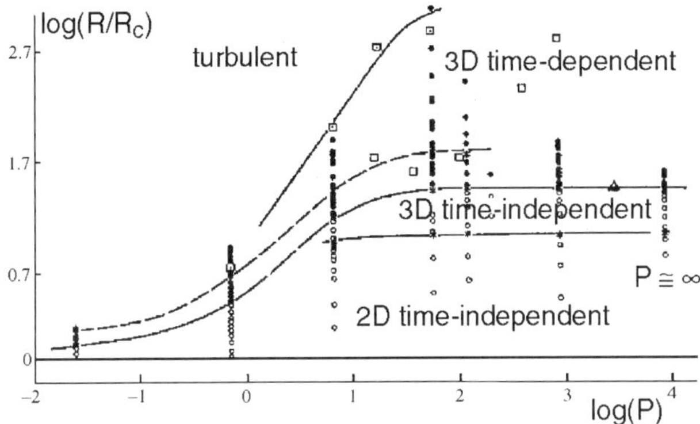

Fig. 3.6. Transition to turbulence in RB convection: experimental results collected before 1973 by Krishnamurti [28].

In all cases the regime called "turbulent convection" was reached after a finite number of steps. At first sight this fact seemed to support the revision by Ruelle and Takens [15] of the classical Landau theory of transition to turbulence [16]: three or four bifurcations before unpredictable behavior instead of n-periodicity resulting from an infinite cascade of Hopf bifurcations. However, the situation was not as satisfactory as one would have liked because the threshold values of the observed transitions were not always well defined and a residual, more or less random, component of the time dependence was most often recorded before it was decided to consider the system as turbulent. This was due to the fact that experiments were performed in wide containers (i.e., with many wavelengths), whereas the transitions were most sensitive to defects in the roll patterns. For example, a state observed at higher R could anticipate transition by nucleating at dislocations. Checking the Ruelle–Takens dynamical systems approach more carefully thus required a better control of the patterns. This has been achieved by playing with confinement effects: only a small number of configurations are available when the number of cells is limited, which makes the Ruelle–Takens approach more relevant a priori, as exemplified in the following section. The case of weakly confined systems is reviewed next.

---

[2] That is, local functions of only one horizontal coordinate, say x, in addition to z.

3.4 Weak Turbulence in Confined Systems and Chaos

When the lateral size of the container is of the order of its height, hence the aspect ratio defined as $\Gamma = \ell/h$ (where ℓ is some typical horizontal scale) is of order one, the structures of the convection modes strongly depend on its geometry. The flow field can be analyzed in terms of a small number of elementary motions characterizing the recirculating cells and to which individual amplitudes can be attached. This directly leads to an interpretation of the observed dynamics in terms of couplings between these variables. Formally, the success of this approach rests on the appropriateness of the strategy of *reduction to the center manifold* that expresses Haken's *slaving principle* [17] in mathematical terms. Basically, this "principle" says that, among the infinitely many degrees of freedom accounting for a continuous medium, most can be eliminated owing to dissipation that smooths out all high-frequency small-scale motions to leave but few slowly evolving fundamental modes. This fundamental property opens the way to an abstract analysis using the whole vocabulary and techniques of dynamical systems theory. In practice, the reduction is, however, analytically out of reach, so that the understanding gained might sound like a conjecture about how things really happen. The beauty and the strength of that conjecture rest on the applicability of the concepts brought forward by the mathematical theory, especially their universal contents.

In fact, RB convection has concentrated a large part of the efforts in the study of nontrivial features of nonlinear dynamics as applied to physical problems, namely, chaos, transition scenarios, strange attractors, and the empirical reconstruction of experimental nonlinear dynamics. The main routes to chaos predicted by theory have been observed: the subharmonic cascade [18], the two-periodic route and its frequency lockings [19], several types of intermittency, and even less generic situations such as quasi-periodic regimes with four or five frequencies. It turns out that, although one is unable to predict which scenario will take place in a given situation, when the system is engaged in a given well-identified route, it strictly follows that route in its most intricate mathematical properties until unavoidable experimental limitations enter to blur the details. Here is the example of type III intermittency [20], the intermittency that develops beyond a subcritical subharmonic bifurcation. This scenario can be modeled by means of an iteration [21]:

$$X_{n+1} = -(1 + r)X_n - X_n^3,$$

where X represents the amplitude of the departure from a limit cycle (corresponding to the fixed point of the iteration at $X = 0$) and r is a control parameter (negative below threshold, positive above). This local map has to be completed by a global assumption about the nature of the manifold on which this reduced dynamics takes place and regarding the return of escaping iterates in the vicinity of the fixed point. Type III intermittency was observed by Dubois et al. [20] in convection using silicone oil (large P). A typical time series is displayed in Figure 3.7 (top). Displayed on the bottom line of that figure, the return maps of the maxima I_n of some observable plotted every two determinations, that is, I_{n+2} as a function of I_n and the statistics of the durations of closely periodic sequences before escape—the so-called laminar intermissions—both agree quantitatively with corresponding theoretical predictions after appropriate empirical rescaling.

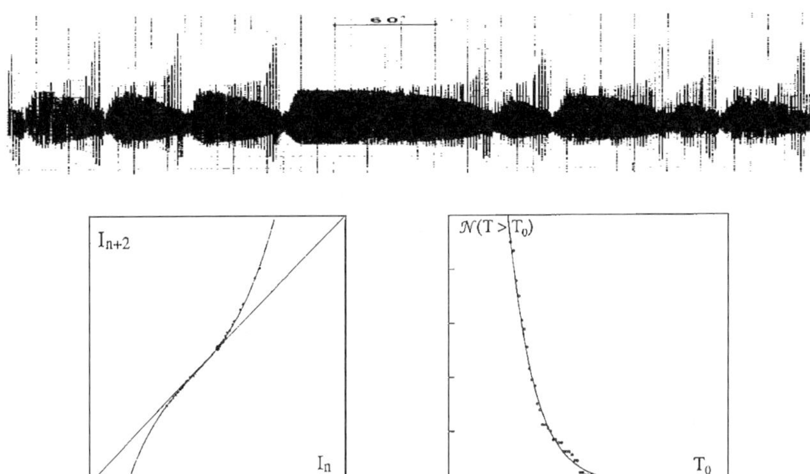

Fig. 3.7. Type III intermittency in RB convection after Dubois et al. [20]. Top: time series of a velocity component measured at some point in the cell. Bottom-left: effective iteration obtained by displaying maxima of that variable every two steps. Bottom-right: cumulative distribution function of the duration of laminar intermissions.

In the same spirit, Takens' *method of delays* [22] has been used extensively to reconstruct attractors [19] and determine quantities such as fractal dimensions and Lyapunov exponents. An early example is shown in Figure 3.8, again taken from the work of the Saclay group [23]. There, the correlation dimension [3] [24] has been determined from the so-called correlation integral (left) computed with reconstructions at ever-larger embedding dimensions d_e and the saturation observed (right) for $d_2 = 2.8$ clearly indicates the low dimensionality of the corresponding chaotic attractor.

These approaches in terms of dynamical systems with few degrees of freedom culminated with the study of the universality and the multifractal properties of chaos emerging from a two-periodic regime observed by Jensen et al. [25] in a

---

[3] This quantity is often used to characterize the fractal component of strange attractors. It is equivalent to the Renyi dimension d_2. The Renyi dimensions d_q are defined by $d_q = \lim_{\epsilon \to 0} \log(\sum_i p_i^q) / \log(\epsilon)$, where ϵ measure the size of balls covering the fractal set and p_i the occupation probability of ball i belonging to the covering of the set. From their usual definitions, it is easily shown that the fractal dimension (or capacity), the information dimension, and the correlation dimension are obtained as d_q with $q = 0, 1,$ and 2, respectively.

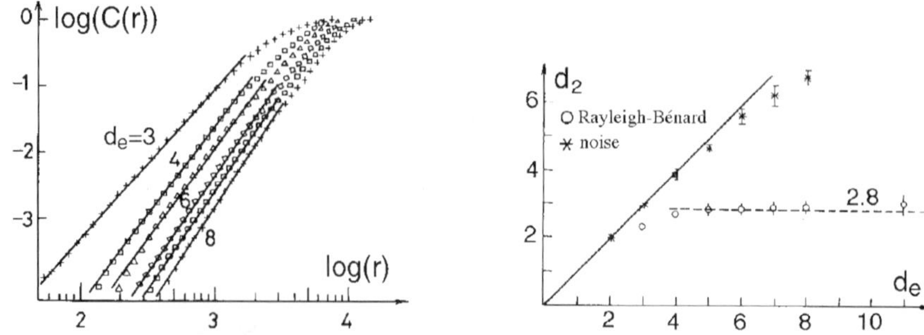

Fig. 3.8. Determination of fractal dimensions by Malraison et al. from Saclay RB convection data. Left: Grassberger–Procaccia correlation integral $C(r)$ as a function of the distance r in embedding spaces with dimensions d_e ranging from 3 to 8. Right: correlation dimension (slope of $C(r)$ in log–log coordinates) for the RB experiment saturating at $\nu = 2.8$ (for comparison: increase observed for a synthetic noise signal). After [23].

convection experiment with a conducting fluid (mercury) under periodic forcing by a variable magnetic field. The left part of Figure 3.9 displays a section of the two-periodic attractor at the margin of chaos and its right part the corresponding "f-of-alpha spectrum" characterizing the distribution of singularities of the distribution of points along the section.

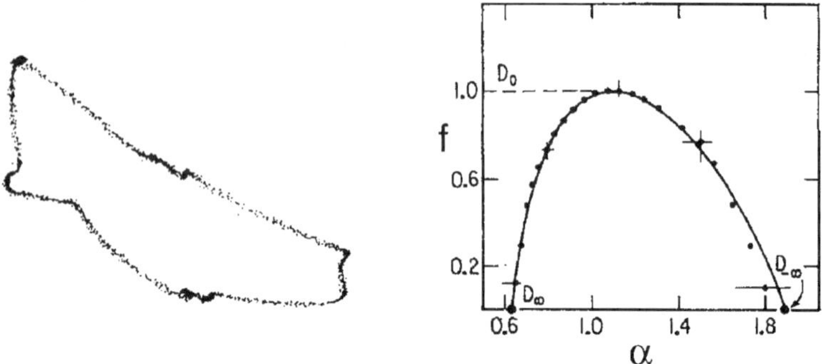

Fig. 3.9. Multifractal properties of the attractor at the threshold of the two-periodic route to chaos after Jensen et al. [25]. Left: section of the attractor. Right: f-of-alpha spectrum.

It should again be stressed that the understanding of the transition to temporal chaos has not been obtained *ab initio* from the primitive problem, for example, by truncating a Galerkin expansion as done to obtain the celebrated Lorenz model [26] (or its higher-dimensional generalizations), which is analytically tractable only at the price of nonphysical boundary conditions (stress-free at top and bottom, horizontally periodic). Rather, an *inductive* type of modeling has been developed, resting on mathematical properties with universal contents in the dynamical systems sense, that is, normal forms for bifurcations and their consequences. This phenomenological approach involving generic dynamical systems, especially iterations, led to impressive results on a *local* scale in phase space but, as a matter of fact, beyond threshold the *global* phase-space structure of a confined RB convection system with realistic boundary conditions becomes extremely complicated. This holds true in particular for large-P fluids at high R, in which thin fluctuating internal thermal boundary layers were observed as illustrated in Figure 3.10, whereas temporal chaos concepts should remain relevant to the dynamics.

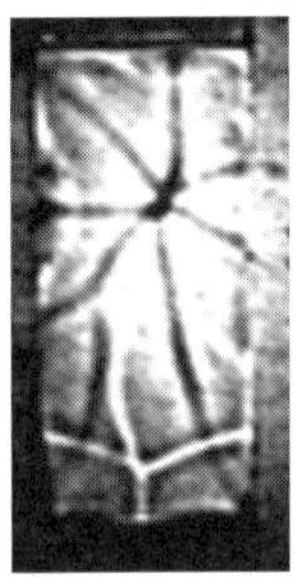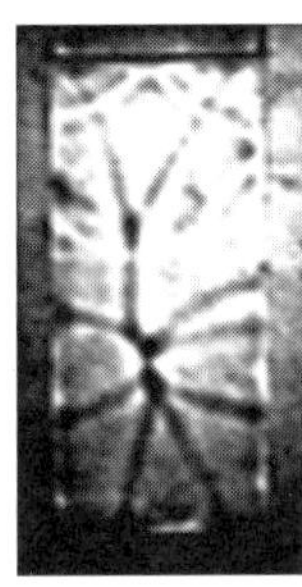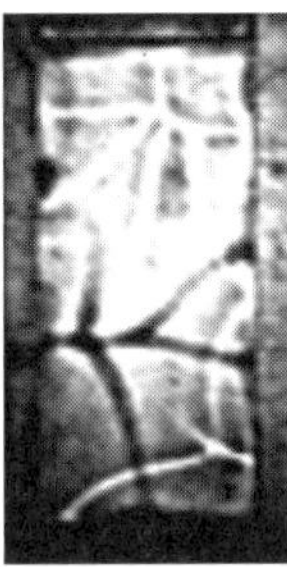

Fig. 3.10. Complicated patterns in silicone oil at high R. Adapted from [28], courtesy of F. Daviaud (Saclay).

3.5 Patterns in Extended Systems at Large P

Furthermore, coexistence of separate attraction basins for different scenarios appeared to be the rule. For example, Type I intermittency and the two-periodic route were observed in the same experimental container with the same fluid (silicone oil) but starting with initial conditions built according to different protocols [27].

In the case of confined systems, the validity of universality concepts is backed by the reduction to low-dimensional dynamical systems through adiabatic elimination of slaved variables. Appropriate adaptations are clearly needed for weakly confined systems. On general grounds, lateral confinement effects are expected to scale as $1/\ell^2$. This turns out to be an advantage because, for sufficiently wide systems ($\Gamma \gg 1$), interesting phenomena may happen in a narrow neighborhood of the threshold, thus accessible to perturbation techniques. The new meaning of universality for structures with many cells can indeed be approached in terms of modulations brought to uniformly periodic roll systems.

The standard multiscale formalism [8] is the most natural framework for the study of weakly disordered patterns. Assuming that, close to the threshold, a modulated solution locally condensed on a pair of wavevectors $\mathbf{k}_{\pm} = \pm k_{\mathrm{c}}\hat{\mathbf{x}}$ can be searched for in the form

$$V(x, y, z, t) = A(x, y, t)\overline{V}(z)\exp(ik_{\mathrm{c}}x) + \text{c.c.}$$

A being a slowly varying envelope ($\partial_t \sim r$, $\partial_x \sim r^{1/2}$, $\partial_y \sim r^{1/4}$, $r = (R - R_{\mathrm{c}})/R_{\mathrm{c}} \ll 1$). One is led to the Newell–Whitehead–Segel (NWS) equation [29, 30]:

$$\tau_0 \partial_t A = \left(r - g|A|^2\right) A + \xi_0^2 \left[\partial_x + \frac{1}{2ik_{\mathrm{c}}}\partial_{yy}\right]^2 A,$$

where τ_0 is the natural relaxation time of fluctuation and ξ_0, linked to the curvature of the marginal stability curve at threshold, is the natural coherence length accounting for the reluctance of the system to adopt a wavevector with a value different from k_{c}.

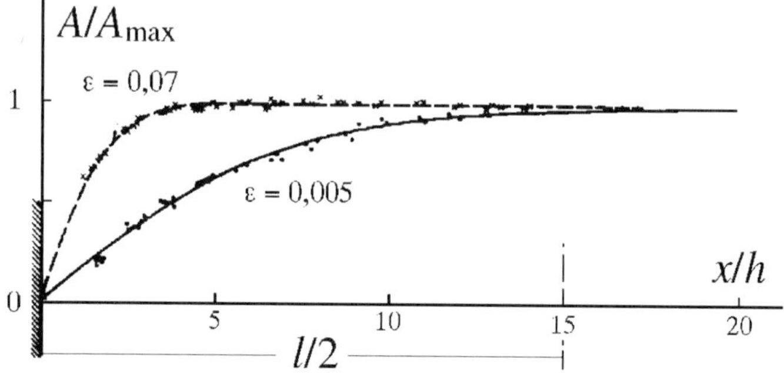

Fig. 3.11. The amplitude of rolls parallel to a lateral wall follows the hyperbolic-tangent law predicted by the Ginzburg–Landau theory up to the quantitative level according to the measurements reported in [31].

This equation was obtained for stress-free boundary conditions but can be shown to hold in the no-slip case with coefficients that can be computed and checked against experiments. As seen in Figure 3.11, the modulation of the convection amplitude close to a lateral boundary parallel to the roll axis predicted by the theory is in excellent agreement with that experimentally measured by Wesfreid et al. [31] in a long and narrow parallelepipedic container using high-P silicone oil. An equally good quantitative agreement was observed for other physical properties, for example, the relaxation rate of fluctuations (critical slowing down), showing that the classical Landau theory of second-order phase transitions applies also to supercritical bifurcations in extended media.[4]

In the vicinity of the convection threshold, a large body of results about the dynamics of disordered patterns, also called *textures*, can be understood within

---

[4] Corrections to the classical behavior due to fluctuation effects similar to those in thermodynamic phase transitions are not detectable in most practical situations. These effects have been recently studied by Oh and Ahlers [32].

the envelope formalism [33]. This remains true as long as the textures behave relaxationally, which is the case when $P \gg 1$. A single scalar envelope field can then be used, that is locally governed by the NWS equation. Owing to its real coefficients, this equation derives from a potential so that solutions to it relax towards essentially time-independent states. Solutions may be disordered, with curvature and defects, owing to the fact that they have to accommodate contradictory requirements, the most important ones being that the roll axis be perpendicular to the lateral boundaries and that the local wavelength be roughly constant and equal to its critical value in the bulk. Even in the limit $P \to \infty$, the envelope formalism is difficult to handle when textures are strongly disordered [34]. An alternative is recourse to simplified models such as the Swift–Hohenberg (SH) model [15]:

Fig. 3.12. Time-independent textured state obtained by simulation of the SH model for $r = 0.3$ in a large square domain with boundary conditions $w = \nabla_\perp w = 0$ where $\nabla_\perp$ is the gradient along the normal to the boundary ($k_c = 1$, $\ell = 200$).

$$\tau_0 \partial_t w = rw - \xi^4 (\nabla^2 + k_c^2)^2 w - g\,w^3,$$

where w is now a two-dimensional field function of time and the horizontal coordinates representing the local convection motion (e.g., the temperature at midheight). The SH model was obtained by a semi-rigorous elimination of the vertical dependence through a Galerkin expansion of the thermohydrodynamic fields in the stress-free case. The original SH model, with its cubic nonlinearity, derives from a potential and, as such, can only lead to time-independent textures. A typical simulation result obtained by myself is displayed in Figure 3.12, illustrating the frustration of geometrical origin with "grains" of well-oriented rolls and several kinds of defects joining them. Nonvariational corrections can be expected, however, leading to unsteady textures with very slow residual time dependence, at least as long as P is sufficiently large. Accordingly, the transition to turbulence can mostly be interpreted in terms of a "fusion" of two-dimensional (scalar) textures. Variants of the SH model have been proposed to deal with convection in different circumstances, when hexagons or squares are expected owing to non-Boussinesq or heat-conductivity effects [36]. When P decreases, the situation is more complicated because the velocity field recovers its rights to control the dynamics. Turbulence is then seen to occur at a moderate distance from the primary convection threshold (see Figure 3.6). A theoretical digression is, however, necessary before I come back to this problem.

3.6 Weak Turbulence in Extended Systems at Small P

Within the stress-free model, Siggia and Zippelius [37] have shown that the NWS envelope equation must be corrected at the lowest order to account for drift flows induced by the curvature of rolls. They obtained a set of two coupled equations for the amplitude A and the intensity U of the drift flow. The effect of the latter is to push the rolls, hence a modified NWS equation

$$\tau_0(\partial_t + ik_{\rm c}U)A = \left(r - g|A|^2\right)A + \xi_0^2 \left[\partial_x + \frac{1}{2ik_{\rm c}}\partial_{yy}\right]^2 A$$

the drift flow U arising from the large-scale vertical vorticity $\Omega = -\partial_y U$ induced by the curvature of the rolls according to

$$\gamma\partial_t\Omega - \partial_{yy}\Omega = g\,\partial_y \left[A^* \left(\partial_x + \frac{1}{2ik_{\rm c}}\partial_{yy}\right)A + {\rm c.c.}\right],$$

where $g \propto (1+P)/P^2 \sim 1/P$ for $P \ll 1$ and c.c. denotes the complex-conjugate term. By contrast with the NWS equation, the new system does not derive from a potential so that a less trivial dynamics can develop.[5] Systematic expansions in this vein have been developed by Decker and Pesch [38] for the realistic case of no-slip boundary conditions leading to an account of stability properties of the rolls in the neighborhood of the threshold in agreement with Busse's previous numerical findings [7] (and analytical results by Piquemal and myself for the zigzag instability [39]).

Close to the threshold, but not asymptotically close to it, the stability of well-aligned patterns can be studied within the phase formalism [40]. Technically the full solution, symbolically written $V(x, y, t)$, is searched for in the form $V_0[k_0\,x + \phi(x, y, t)] + V_1 + ...$, where ϕ is a slowly varying phase and x the coordinate along the wavevector of the reference roll system. A compatibility condition for ϕ leads to a diffusion equation:

$$\partial_t\phi = D_{\parallel}\partial_{xx}\phi + D_{\perp}\partial_{yy}\phi,$$

---

[5] The Galilean invariance of the stress-free Boussinesq problem is broken when no-slip boundary conditions are considered but, when P is sufficiently small, this invariance is approximately restored beyond threshold through thin viscous boundary layers at the top and bottom plates so that the approach still makes sense.

where the diffusion coefficients $D_\parallel$ and $D_\perp$ are functions of the wavevector k_0 different from $k_{\rm c}$ (but not too far from it) and the relative distance to threshold r. Universal secondary instabilities occur when these diffusion coefficients become negative (Eckhaus when $D_\parallel < 0$; zigzags when $D_\perp < 0$). The generalization of this approach is due to Newell and Cross [34] who derive the diffusion equation in a form independent of a reference frame linked to the local roll wavevector. The solution being now written as $V_0(\Theta(x,y,t)) = A(x,y,t)\exp(i\Theta(x,y,t))$, the local wavevector is given by $\mathbf{k}_{\rm h}(x,y,t) = \boldsymbol{\nabla}_{\rm h}\Theta(x,y,t)$, $k \equiv \sqrt{\mathbf{k}_{\rm h}^2}$, the amplitude A is enslaved to the phase Θ through an eikonal equation $A = A(k)$, and Θ is governed by

$$\tau(k)\partial_t\Theta(x,y,t) + \boldsymbol{\nabla}_{\rm h} \cdot [\mathbf{k}_{\rm h}B(k)] = 0,$$

where $\tau(k)$ and $B(k)$ are two functions of k. The previous diffusion equation is recovered when a nearly uniform roll pattern is assumed, which yields $D_\parallel = -\tau^{-1}{\rm d}(kB)/{\rm d}k$ and $D_\perp = -\tau^{-1}{\rm d}B/{\rm d}k$. Cross and Newell then added drift flows phenomenologically through the change: $\partial_t \mapsto \partial_t + \mathbf{U} \cdot \boldsymbol{\nabla}_{\rm h}$. The field $\mathbf{U}$ that generalizes the variable U introduced earlier now describes a horizontal incompressible flow deriving from a stream function Ψ (i.e., $\mathbf{U} \equiv (\partial_y\Psi, -\partial_x\Psi)$) governed by

$$\nabla^2\Psi = \gamma\,\hat{\mathbf{z}} \cdot \boldsymbol{\nabla}_{\rm h} \times \left[\mathbf{k}_{\rm h}\left(\boldsymbol{\nabla}_{\rm h} \cdot \left(\mathbf{k}_{\rm h}A^2\right)\right)\right],$$

where A is the amplitude and $\hat{\mathbf{z}} = \hat{\mathbf{x}} \times \hat{\mathbf{y}}$. In practice, in the unstable range, generalized phase equations form an ill-posed problem leading to singularities and some regularization is demanded, as discussed in detail by Newell and coworkers (see, e.g., [8]).

These theoretical problems are indeed fully relevant to the understanding of the transition to turbulence in low Prandtl number fluids that do not behave relaxationally but have more active dynamics of inertial origin. Whereas it was known for long from conventional studies that turbulent states can be observed at moderate R when $P \sim 1$ or below (see Figure 3.6), it was later discovered that the range of strictly time-independent convection was extremely narrow at large aspect ratios $\Gamma \gg 1$ [65]. However, the kind of time-dependence that developed was not real high-frequency turbulence but resembled a low-frequency broadband noise with a power-law spectrum appearing before any trace of secondary instability [42]. This behavior was not well understood until Pocheau et al. [43] showed that this noise resulted from a cycle involving the nucleation, migration, and annihilation of dislocations. Pictures of this cycle are displayed in Figure 3.13.

In an attempt to get a semi-microscopic account of convection at low Prandtl number in weakly confined systems, it seems necessary to include drift flow effects in the SH model. Paralleling Siggia and Zippelius, I derived a generalized SH model by truncating at lowest significant order a Galerkin expansion of the thermohydrodynamic fields while taking care of the large-scale flow driven by curvature effects, obtaining [45, a]:

$$\partial_t w + \mathbf{U} \cdot \boldsymbol{\nabla}_{\rm h}w \;=\; \left[r - (\boldsymbol{\nabla}_{\rm h}^2 + 1)^2\right]w - \mathcal{N}(w),$$
$$(\partial_t + P\boldsymbol{\nabla}_{\rm h}^2)\boldsymbol{\nabla}_{\rm h}^2\Psi = \partial_y w\,\partial_x\boldsymbol{\nabla}_{\rm h}^2 w - \partial_x w\,\partial_y\boldsymbol{\nabla}_{\rm h}^2 w,$$

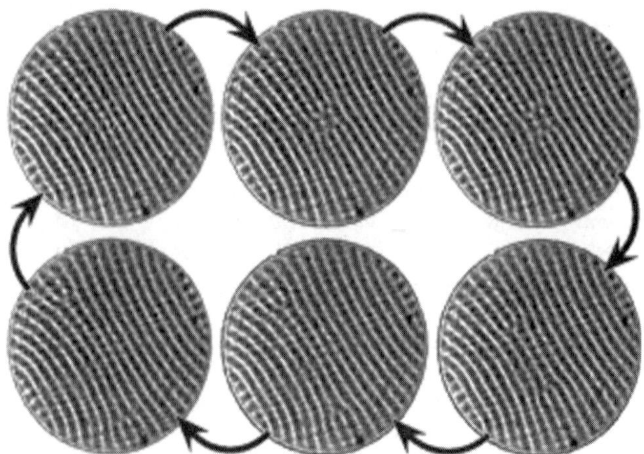

Fig. 3.13. In a cylindrical container (with argon under pressure as the working fluid), lateral boundary effects on the roll orientation at the walls imply some curvature in the buck. Interplaying with drift flow, a periodic process of nucleation, migration, and annihilation of dislocations develops, at first roughly periodic in time. Irregularities in the process lead to a noisy dynamics with power-law spectrum. Pictures kindly provided by V. Croquette (ENS-Ulm); see [25] for details.

where, as above, Ψ is the stream function from which the velocity field $\mathbf{U}$ derives and $\mathcal{N}(w)$ some nonlinear saturating term (either w^3 as in the original SH model, or for example, $[(\boldsymbol{\nabla}_\mathrm{h} w) + w^2]w$ as in [45, a], or possibly more general—even nonlocal—expressions). Numerical simulations of the extended SH model [45, b] have led to an interpretation of weak turbulence in extended RB convection systems in terms of a dynamical compatibility of drift flows to the spatially disordered topology of patterns implied by the geometrical frustration imposed by the lateral boundaries.

At larger but still small P, a quite different kind of weak turbulence, called *spiral–defect chaos* (SDC) has also been observed by Ahlers and coworkers [46] among others (Chapter 4 by Ahlers in this volume). This regime is illustrated in Fig. 3.14 for two different values of R. Bistability with respect to this regime has also to be noted: at the same R, straight rolls are still locally stable but, when it is nucleated somewhere, the SDC regime invades the whole surface of the system at the expense of the rolls. From a modeling viewpoint, the SDC regime has been observed in simplified systems such as the generalized SH model presented above [47] or other simplified models in the same spirit [84]. The role of drift flows seems essential for such a space–time chaotic behavior but, on more general ground, its extensive character (i.e., a density of spiral cores can be defined, independent of the shape of the container provided it is sufficiently wide) makes it interesting from the point of view of a statistical analysis of out-of-equilibrium systems at the thermodynamic limit.

Fig. 3.14. The density of spiral cores increases as R gets farther beyond the threshold of SDC. Pictures kindly provided by G. Ahlers (UCSB); see [46].

3.7 Space–Time Intermittency and Statistical Physics

Another situation is of interest with respect to statistical aspects of the transition to turbulence when confinement effects are exploited to produce a quasi-one-dimensional pattern; see Figure 3.15. In the case described, the fluid is silicone oil with moderate to high Prandtl number so that no strong large-scale flow effects are expected. Seen from above the positions of the thin thermal boundary layers between consecutive cells can be recorded as a function of time to construct space–time diagrams. The most interesting experiments have been performed in the narrow annular cell shown in Figure 3.16 (top), thus avoiding parasitic end effects.

The transition to turbulence in this system happens via a new specific scenario: the time-independent regular arrangement of cells below the transition is disrupted by intermittent chaotic bursts above the threshold. At a given time the system can be divided in laminar and turbulent domains and at a given point in space the system is alternatively laminar or turbulent, hence the term *space–time intermittency* (STI). This kind of transition has been observed in several convection experiments [49, 50] but also in other physical systems. For a review, consult [51].

The theoretical account of this scenario follows Pomeau's idea [53] of an equivalence of STI with a time–oriented stochastic process known as *directed percolation* in statistical physics. This process deals with the modeling of epidemic processes in which subjects in good health (in the so-called absorbing state) are contaminated by some disease (the excited state) with finite local probability. The subjects stand at the nodes of a lattice and contamination is from one node to its neighbors while time is advancing by steps. Above some probability threshold, contamination is sustained and propagates to infinity with finite probability, otherwise the epidemic ceases spontaneously. Directed percolation is a critical phenomenon that defines a so-called universality class, with a specific set of critical exponents, controlling, for example, the fraction of excited states that grows as a power law of the distance to the threshold.

Fig. 3.15. Convection in four parallel slits with decreasing width from (a) to (d), seen from above. The linear arrangement of cells with complex individual flow patterns observed in (a) is similar to those in Figure 3.10. In (d) a regular row of simpler cells has set in, hence a quasi-one-dimensional pattern with fewer local degrees of freedom. Adapted from [28], courtesy of F. Daviaud (Saclay).

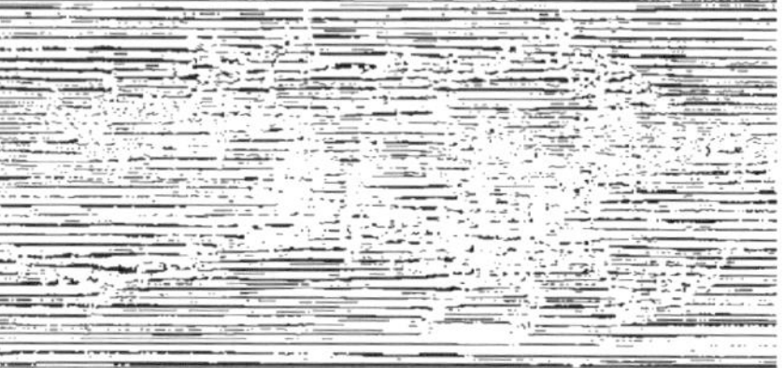

Fig. 3.16. Transition to space–time intermittency in quasi-one-dimensional geometry. Top: annular convection cell, seen from above (the radial aspect ratio is similar to the transverse one in Figure 3.15 (d)). Bottom: space–time diagrams with the azimuthal coordinate along the vertical and time running from left to right, below (left) and above (right) the STI threshold. Courtesy of F. Daviaud (Saclay); see [28] for details.

An essential assumption is that, out of the two possible coexisting states at each location, the one playing the role of the absorbing state is locally stable so that there cannot be spontaneous birth of excited states. Coexistence of states in the local phase space usually manifests itself by the formation of walls for systems distributed in space. In the simplest case of a potential system these walls move so as to minimize the potential. In the more complicated case where the excited state is a chaotic transient, the motion of the wall through contamination becomes random. In order to better understand how STI can occur in deterministic systems, that is to say, how local transient temporal chaos can be converted in sustained space–time chaos by the interaction between neighboring subsystems, models in terms of *coupled map lattices* [54] have been built with an appropriate local phase space structure [55]. The universality issue turns out to be intricate, especially regarding the thermodynamic limit of infinitely large systems in the long-time limit, after transients have decayed [56]. As far as convection in the quasi-one-dimension is concerned, turbulence is obviously the excited state. The statistics of the size of turbulent domains has been studied as a function of R, showing that the STI transition was only imperfect [50] due to a tiny probability of spontaneous nucleation of turbulent cells (possibly linked to the focusing of a long-wavelength secondary instability of the convection cells chain).

3.8 Convection at Large Rayleigh Numbers

Beyond the transitional stage, convection enters a fully developed turbulent regime usually best characterized by physical properties scaling as some power of R, as noticed by Siggia [11]. Most often one is interested in the behavior of the Nusselt number defined as

$$N = \frac{\text{total heat flux}}{\text{conduction heat flux}},$$

where the "total heat flux" is the measured flux and the "conduction heat flux" is the flux computed from the applied temperature difference upon assuming Fourier law in a fluid at rest, hence $N = 1$ below the primary threshold ($R < R_c$), and $N - 1$ measuring the contribution of beyond. In the weakly nonlinear regime close to threshold it is expected to vary as

$$N - 1 \propto v_z \theta \propto \frac{R - R_c}{R_c}$$

because $N - 1$ is the average over the cell of the product θv_z, and each term scales as $[(R - R_c)/R_c]^{1/2}$. This behavior has been well observed experimentally long ago and, as mentioned by Chandrasekhar [1], can serve to locate the threshold with precision. When R becomes larger than a few hundreds of R_c, N increases with R as a power law:

$$N \sim R^\gamma$$

but some confusion exists about the relevant value of γ that may depend on the range of R considered, with crossovers between different regimes, the nature of the fluid (value of P) and to a lesser extent on the shape of the container or the roughness of the top and bottom plates. Exact bounds can be given for γ (see [57] for references) and early arguments predicted $\gamma = 1/3$ or $\gamma = 1/2$. The first value is obtained by assuming that the turbulent heat flux is fully controlled by finite-width thermal boundary layers so that it becomes independent of the container height [12] and the second one, considered as the "ultimate regime" at asymptotically large R, is reached when buoyancy is fully balanced by advection in the momentum equation so that the heat transfer no longer depends on molecular properties κ and ν (Kraichnan, 1962). Consult [58] for an introductory presentation and [11] or [59] for more information. Some early experiments seemed to support $\gamma = 1/3$ whereas others for $P \ll 1$ yielded rather $\gamma = 1/4$, As a matter of fact, these early studies were performed in extended geometry with many convection cells and did not allow testing of sufficiently large ranges of parameters. Since 1985 new experiments specifically focusing on this problem have been developed; see, for example, [60, 61, 62] among many others. In order to reach high values of R without increasing the temperature difference too much

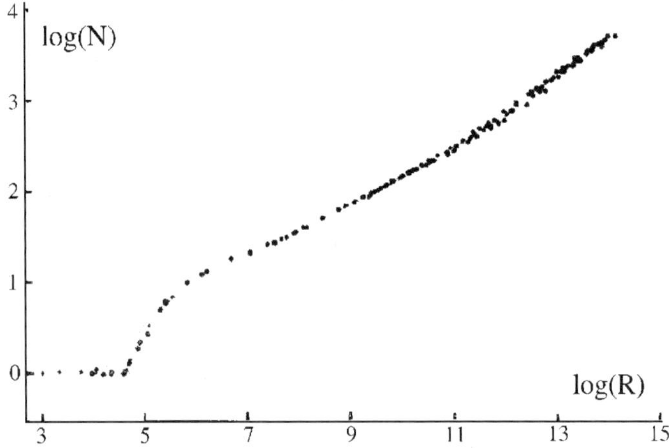

Fig. 3.17. Total heat flux as measured in terms of the Nusselt number as a function of the Rayleigh number (in log–log scale). After the initial abrupt increase corresponding to the weakly nonlinear regime, the Nusselt number varies more gently as R^γ. The exponent is close to 2/7 over a four-decade range between 10^7 and 10^{11}, and somewhat larger, about 1/2, above. Adapted from Chavanne et al. [61, a].

(validity condition for the Boussinesq approximation), containers have indeed to be tall because, from its definition, R grows as h^3 at given ΔT. But large aspect ratios cannot be maintained if the experiment is to stay within reasonable horizontal-size limits. In practice aspect ratios $\Gamma = 1/2$ or 1 have been used and very large R, up to 10^{18} times the critical value, have been achieved. The drawback of small aspect ratios is that the mean large-scale flow may control an important part of the heat transfer, especially by producing a "wind" along the top and bottom plates. For example, thermal transfer through such a created velocity boundary layer yields $N \sim R^{1/4}$ when they are laminar and $N \sim R^{2/7}$ when they are turbulent.

Figure 3.17, adapted from the results of Chavanne et al. in liquid helium [61], is typical of the most recent experiments. One can easily identify the previously mentioned weakly nonlinear regime close to , [6] then the "soft turbulence" regime where chaos is still mostly temporal, next "hard turbulence" with an exponent $\gamma \simeq 2/7$ explained by the theory involving thermal transfer through turbulent layers sheared by the general circulation wind, and finally the "ultimate" regime with exponent tending to $1/2$. The existence of this last regime has been challenged [60] and conditions for its observation clarified by Roche et al. [61, c].

The different turbulence regimes have been reconsidered by Grossmann and Lohse [59] who distinguish them from the origin of the main contributions to the dissipation, in the bulk or within boundary layers, and give a unified scaling picture of the strong turbulence problem in RB convection. According to this picture, four main regimes can exist with different γ signatures and it is argued that the scaling with $\gamma = 2/7$, although observed over a rather wide range of Rayleigh numbers (and in spite of its appealing physical interpretation in terms of heat transfer through turbulent boundary layers), is equally well accounted for as a blend of regimes with $\gamma = 1/4$ and $\gamma = 1/3$ with appropriate weighting coefficients. How these regimes depend on the Prandtl number (also possibly on the shape of the container and associated bifurcations of the mean large-scale flow) is the subject of recent studies, for example, [62].

3.9 Conclusion

One century after his first experiments, one cannot but notice that Bénard opened a particularly rich field of research, accessible to detailed analysis both theoretical and experimental. Once the initial misunderstanding about the role of surface tension has been cleared up, RB convection (within the Boussinesq approximation) has indeed presented itself as an ideal testing ground for the interplay of mathematics and physics methods, especially during the last thirty years. In particular, fundamental problems related to universality could be tackled, both for confine systems where the theory of dynamical systems is relevant (chaos and transition scenarios, e.g., the subharmonic cascade) and for extended systems where statistical physics is an appealing framework (Ginzburg–Landau formalism and nonlinear pattern selection, space–time intermittency and directed percolation). It should further be noticed that progress has been obtained through an exemplary feedback process involving experiments, theory, and modeling. Let us hope that the methods developed to reach such an improved un-

---

[6] The threshold is shifted somewhat beyond 1708 owing to small aspect ratio effects.

derstanding of the emergence of complexity in this specific physical system will also fuel the study of issues crucial to the future of our natural environment.

References

1. H. Bénard, "Les tourbillons cellulaires dans une nappe liquide," *Rev. Gén. Sci. pures et appl.* **11** (1900) 1261–1271, 1309–1328.
2. Lord Rayleigh, "On the convective currents in a horizontal layer of fluid when the higher temperature is on the under side," *Phil. Mag.* **32** (1916) 529–546.
3. J.R.A. Pearson, "On convection cells induced by surface tension," *J. Fluid Mech.* **4** (1958) 489–500.
4. P. Cerisier, C. Jamond, J. Pantaloni, J.C. Charmet, "Déformation de la surface libre en convection de Bénard–Marangoni," *J. Physique* **45** (1984) 405–411.
5. S. Chandrasekhar, *Hydrodynamic and Hydromagnetic Stability* (Clarendon, Oxford, 1961).
6. E.L. Koschmieder, *Bénard Cells and Taylor Vortices* (Cambridge University Press, UK, 1993).
7. F.H. Busse, "Non-linear properties of thermal convection," *Rep. Prog. Phys.* **41** (1978) 1929–1967, and "Transition to turbulence in thermal convection ," in *Convective Transport and Instability Phenomena*, J. Zierep, H. Oertel, Jr., eds. (Braun, Karlsruhe, 1982).
8. A.C. Newell, Th. Passot, J. Lega, "Order parameter equations for patterns," *Ann. Rev. Fluid Mech.* **25** (1993) 399–453.
9. M.C. Cross, P.C. Hohenberg, "Pattern formation outside of equilibrium," *Rev. Mod. Phys.* **65** (1993) 851–1112.
10. E. Bodenschatz, W. Pesch, G. Ahlers, "Recent developments in Rayleigh–Bénard convection," *Ann. Rev. Fluid Mech.* **32** (2000) 708–778.
11. E.D. Siggia, "High Rayleigh number convection," *Ann. Rev. Fluid Mech.* **26** (1994) 137–168.
12. W.V.R. Malkus, G. Veronis, "Finite amplitude cellular convection," *J. Fluid Mech.* **38** (1958) 227–260.
13. A. Schlüter, D. Lortz, F.H. Busse, "On the stability of finite–amplitude convection," *J. Fluid Mech.* **23** (1965) 129–144.
14. R. Krishnamurti, "Some further studies on the transition to turbulent convection," *J. Fluid Mech.* **60** (1973) 285–303.
15. D. Ruelle, F. Takens, "On the nature of turbulence," *Commun. Math. Phys.* **20** (1971) 167–192. Addendum **23** (1971) 344.
16. L.D. Landau, "On the problem of turbulence," reprinted in *Collected Papers of L.D. Landau*, D. ter Haar, ed. (Pergamon, Oxford, 1965) pp. 387–391.
17. H. Haken, *Synergetics* (Springer, 1983).
18. A. Libchaber, J. Maurer, "Une expérience de Rayleigh–Bénard en gémométire réduite; multiplication, accrochage et démultiplication de fréquences," *J. Physique Colloques* **41–C3** (1980) 51–56.
19. M. Sano, Y. Sawada, "Experimental study on Poincaré mappings in Rayleigh–Bénard convection" in *Turbulence and Chaotic Phenomena in Fluids*, T. Tatsumi, ed, (Elsevier, New York 1984).
20. M. Dubois, M.A. Rubio, P. Bergé, "Experimental evidence of intermittencies associated with a subharmonic bifurcation," *Phys. Rev. Lett.* **51** (1983) 1446–1449.

21. Y. Pomeau, P. Manneville, "Intermittent transition to turbulence in dissipative dynamical systems," *Commun. Math. Phys.* **74** (1980) 189–197.
22. F. Takens, "Detecting strange attractors in turbulence," in *Dynamical Systems and Turbulence*, D.A. Rand, L.S. Young, eds., Lecture Notes in Mathematics **898** (Springer, New York, 1981) pp. 366–381.
23. B. Malraison, P. Atten, P. Bergé, M. Dubois, "Dimension d'attracteurs étranges: une détermination expérimentale en régime chaotique de deux systèmes convectifs," *C.R. Acad. Sc. Paris* **297** Série II (1983) 209–214.
24. P. Grassberger, I. Procaccia, "Measuring the strangeness of strange attractors," *Physica D* **9** (1983) 189–208.
25. M.H. Jensen, L.P. Kadanoff, A. Libchaber, I. Procaccia, J. Stavans, "Global universality at the onset of chaos: Results of a forced Rayleigh–Bénard experiment," *Phys. Rev. Lett.* **55** (1985) 2798–2801.
26. E.N. Lorenz, "Deterministic nonperiodic flow," *J. Atmos. Sci.* **20** (1963) 130–141.
27. M. Dubois, P. Bergé, "Étude expérimentale des transitions vers le chaos en convection de Rayleigh–Bénard," in *Le Chaos, Théorie et expériences*, P. Bergé, ed. (Eyrolles, 1988) pp. 1–72 (in French).
28. P. Bergé, "From temporal chaos towards spatial effects," *Nuclear Physics B* (Proc. Suppl.) **2** (1987) 247-258.
29. L.A. Segel, "Distant side-walls cause slow amplitude modulation of cellular convection," *J. Fluid Mech.* **38** (1969) 203–224.
30. A.C. Newell, J.A. Whitehead, "Finite bandwidth, finite amplitude convection," *J. Fluid Mech.* **38** (1969) 279–303.
31. J.E. Wesfreid, Y. Pomeau, M. Dubois, Ch. Normand, P. Bergé, "Critical effects in Rayleigh-Bénard convection," *J. Physique* **39** (1978) 725–731.
32. J. Oh & G. Ahlers, "Fluctations in a fluid with a thermal gradient," UCSB preprint, September 2002.
33. M.C. Cross, "Ingredients of a theory of convective textures close to onset," *Phys. Rev. A* **25** (1982) 1065–1076.
34. M.C. Cross, A.C. Newell, "Convection patterns in large aspect ratio systems," *Physica D* **10** (1984) 299–328.
35. J. Swift, P.C. Hohenberg, "Hydrodynamic fluctuations at the convective instability," *Phys. Rev. A* **15** (1977) 319–328.
36. C. Perez-Garcia, J. Millán-Rodriguez, H. Herrero, "A generalized Swif-Hohenberg model for several convective problems," in *Instabilities and Nonequilibrium Structures IV*, E. Tirapegui, W. Zeller, eds. (Kluwer, Hingham, MA, 1993) pp. 225–234.
37. E.D. Siggia, A. Zippelius, "Pattern selection in Rayleigh–Bénard convection near threshold," *Phys. Rev. Lett.* **47** (1981) 835–838.
38. W. Decker, W. Pesch, "Order parameter and amplitude equations for the Rayleigh–Bénard convection," *J. Phys. II France* **4** (1994) 419–438.
39. P. Manneville, J.M. Piquemal, "Zigzag instability and axisymmetric rolls in Rayleigh–Bénard convection, the effects of curvature," *Phys. Rev. A* **28** (1983) 1774–1790.
40. Y. Pomeau, P. Manneville, "Stability and fluctuations of a spatially periodic convective flow," *J. Physique Lettres* **40** (1979) L-610.
41. G. Ahlers, R.P. Behringer, "Evolution of turbulence from the Rayleigh–Bénard instability," *Phys. Rev. Lett.* **40** (1978) 712–716.
42. A. Libchaber, J. Maurer, "Local probe in a Rayleigh–Bénard experiment in liquid Helium," *J. Physique-Lettres* **39** (1978) L369–L372.
43. A. Pocheau, V. Croquette, P. Le Gal, "Turbulence in a cylindrical container of Argon near threshold of convection," *Phys. Rev. Lett.* **55** (1985) 1094–1097.

44. V. Croquette, "Convective pattern dynamics at low Prandtl number. Part I & II." *Contemporary Physics* **30** (1989) 113–133, 153–171.

45. P. Manneville, (a) "A two-dimensional model for three-dimensional convective patterns in wide containers," *J. Physique* **44** (1983) 759–765. (b) "Towards an understanding of weak turbulence close to the convection threshold in large aspect ratio systems," *J. Physique Lettres* **44** (1983) L903–L916.

46. S.W. Morris, E. Bodenschatz, D.S. Cannell, G. Ahlers, "The spatio-temporal structure of spiral-defect chaos," *Physica D* **97** (1996) 164–179.

47. (a) H.-w. Xi, J.D. Gunton, J. Viñals, " Spiral defect chaos in a model of Rayleigh–Bénard convection," *Phys. Rev. Lett.* **71** (1993) 2030–2033. (b) H.-W. Xi, J.D. Gunton, J. Viñals, "Study of spiral pattern formation in Rayleigh–Bénard convection," *Phys. Rev. E* **47** (1993) R2987–R2990. (c) M. Bestehorn, M. Neufeld, R. Friedrich, H. Haken, "Comment on 'Spiral-pattern formation in Rayeigh–Bénard convection'," *Phys. Rev. E* **50** (1994) 625-626.(d) X.-J. Li, H.-W. Xi, J.D. Gunton, "Nature of the roll to spiral-defect-chaos transition," *Phys. Rev. E* **57** (1998) 1705.

48. W. Decker, W. Pesch, A. Weber, "Spiral defect chaos in Rayleigh–Bénard convection," *Phys. Rev. Lett.* **73** (1994) 648–651.

49. S. Ciliberto, P. Bigazzi, "Spatiotemporal intermittency in Rayleigh-Bénard convection," *Phys. Rev. Lett.* **60** (1988) 286–289.

50. F. Daviaud, M. Dubois, P. Bergé, "Spatio-temporal intermittency in quasi one-dimensional Rayleigh-Bénard convection," *Europhys. Lett.* **9** (1989) 441–446.

51. F. Daviaud, "Experiments in 1D turbulence," in [52].

52. P. Tabeling, O. Cardoso, eds., *Turbulence. A Tentative Dictionary* (Plenum, New York, 1994).

53. Y. Pomeau, "Front motion, metastability and subcritical bifurcations in hydrodynamics," *Physica D* **23** (1986) 3–11.

54. For general information consult: K. Kaneko, *Theory and Application of Coupled Map Lattices* (Wiley, New York, 1994).

55. (a) H. Chaté, P. Manneville, "Spatiotemporal intermittency in coupled map lattices," *Physica D* **32** (1988) 409–422. For a review, see (b) H. Chaté, P. Manneville, "Spatiotemporal intermittency," in [52].

56. (a) P. Grassberger, T. Schreiber, "Phase transitions in coupled map lattices," *Physica D* **50** (1991) 177–188. (b) G. Rousseau, B. Giorgini, R. Livi, H. Chaté, "Dynamical phases in a cellular automaton model for epidemic propagation," *Physica D* **103** (1997) 554–563.

57. (a) L. Howard, "Bounds on flow quantities," *Ann. Rev. Fluid Mech.* **4** (1972) 473–494. (b) F.H. Busse, "The optimum theory of turbulence," *Adv. Appl. Mech.* **18** (1978) 77–121. (c) C.R. Doering, P. Constantin, "Variational bounds on energy dissipation in incompressible flows. III. Convection," *Phys. Rev. E* **53** (1996) 5957–5981.

58. J. Sommeria, "The elusive 'ultimate state' of thermal convection," *Nature* **398** (1999) 294–295 (news and views).

59. S. Grossmann, D. Lohse, (a) "Scaling in thermal convection: A unifying theory," *J. Fluid Mech.* **407** (2000) 27–56. (b) "Thermal convection for large Prandtl numbers," *Phys. Rev. Lett.* **86** (2001) 3316–3319.

60. J.J. Niemela, L. Skrbek, K.R. Sreenivasan, R.J. Donnelly, "Turbulent convection at very high Rayleigh numbers," *Nature* **404** (2000) 837–840, err. **406** (2000) 439.

61. (a) X. Chavanne, F. Chillà, B. Castaing, B. Hébral, B. Chabaud, J. Chaussy, "Observation of the ultimate regime in Rayleigh–Bénard convection," *Phys. Rev. Lett.* **79** (1997) 3648–3651. (b) X. Chavanne, F. Chillà, B. Chabaud, B. Castaing,

B. Hébral, "Turbulent Rayleigh–Bénard convection in gaseous and liquid He," *Phys. Fluids* **13** (2001) 1300–1320. (c) P.-E. Roche, B. Castaing, B. Chabaud, B. Hébral, "Observation of the 1/2 power law in Rayleigh–Bénard convection," *Phys. Rev. E* **63** 045303-1-4 (Rapid Communications).

62. G. Ahlers, X. Xu, "Prandtl-number dependence of heat transport in turbulent Rayleigh–Bénard convection," *Phys. Rev. Lett.* **86** (2001) 3320–3323.

4 Experiments with Rayleigh–Bénard Convection

Guenter Ahlers

Department of Physics and IQUEST, University of California
Santa Barbara CA 93106 USA
e-mail: guenter@physics.ucsb.edu

After a brief review in the introduction of the major breakthroughs in the study of Rayleigh–Bénard convection (RBC) since the experiments of Henri Bénard, a few selected topics are presented in more detail. The effect of on the bifurcation to convection is discussed because experimental work on this is quite recent and as yet incomplete. Examples of spatio–temporal chaos are examined because this interesting nonlinear state is as yet incompletely understood. The effect of rotation on RBC is presented because some of the experimental results disagree with modern theories.

4.1 Introduction

Convection in a shallow horizontal layer of a fluid heated from below had been observed on several occasions during the nineteenth century [1]. However, the carefully controlled and quantitative laboratory experiments of Henri Bénard [2] focused the interest of other scientists on this fascinating problem. Bénard studied the patterns of the convective flow in the presence of a free upper surface, using a variety of fluids with different viscosities. He made quantitative determinations of the deformation of the upper surface, of the characteristic length scales of the pattern, and of the direction of flow within the fluid. Although we now know that the beautiful hexagonal patterns observed by Bénard [3] were caused by the contribution of a temperature dependent surface tension, these experiments were the direct motivation of Lord Rayleigh's seminal stability analysis [4] for the case of free horizontal boundaries in the absence of surface tension. Rayleigh's opening remark in his paper in *The London, Edinburgh, and Dublin Philosophical Magazine and Journal of Science* was "The present is an attempt to examine how far the interesting results obtained by Bénard in his careful and skillful experiments can be explained theoretically". Lord Rayleigh recognized that there is a finite value of the temperature difference $\Delta T = \Delta T_c$ for the onset of convection, and that the important combination of parameters that determines the onset is

$$R = \frac{\alpha g d^3 \Delta T}{\kappa \nu}, \tag{4.1}$$

where α is the isobaric thermal expansion coefficient, κ the thermal diffusivity, ν the kinematic viscosity, d the spacing between the plates, g the acceleration of

68   Guenter Ahlers

gravity, and ΔT the temperature difference. We now refer to R as the "Rayleigh number". Lord Rayleigh also found that the instability occurs at finite wavenumber k_c, and that it is a stationary instability (i.e., that the relevant eigenvalues are real). For the free boundary conditions that he used he was able to obtain the analytic results $R_c = 27\pi^4/4$ and $k_c = \pi/\sqrt{2}$.

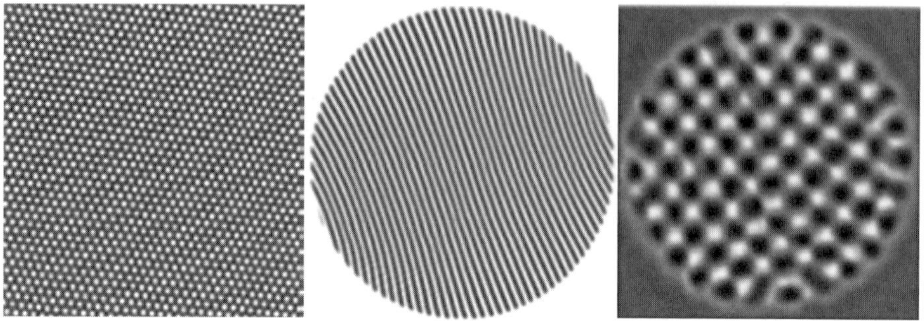

Fig. 4.1. Left: hexagonal pattern of non-Boussinesq convection in compressed SF_6 near its critical point (from [6]). Middle: roll pattern for a Boussinesq fluid (from [7]). Right: square pattern in binary-mixture convection (from [8]).

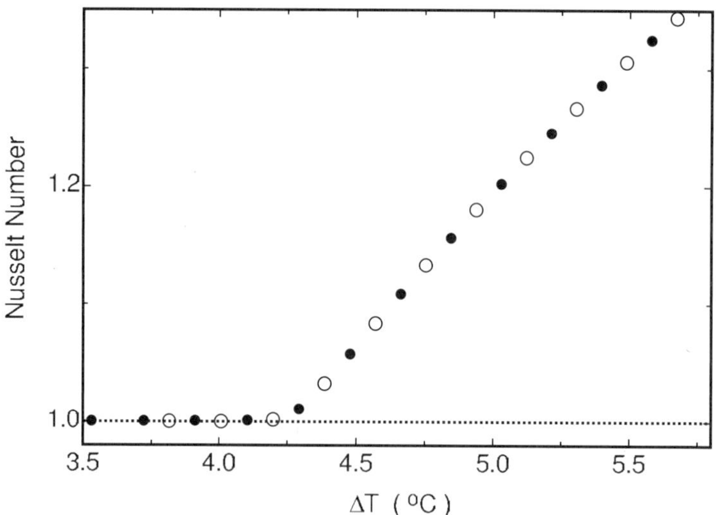

Fig. 4.2. Nusselt number measurements using ethanol in a circular cell with $d = 1.54$ mm and diameter $D = 88$ mm. Open (closed) circles: increasing (decreasing) ΔT (from [14]).

The problem caught the attention of other giants in the field during the next several decades. Here I mention only a few highlights. Rayleigh's work was followed by the stability analysis for more realistic *rigid* boundaries by Sir Harold Jeffreys [5], that (after some numerical problems) yielded the values $R_c = 1708$ and $k_c = 3.117$ relevant to experiments using fluids confined between well-conducting solid parallel plates. There were a number of other milestones. Particularly noteworthy on the theoretical side were the first weakly nonlinear analyses that led to predictions of the stable convection patterns. On the basis of the linear stability analysis of Lord Rayleigh or Sir Harold Jeffreys one knows the magnitude of the critical wavedirector, but one cannot decide whether the patterns above onset will consist of rolls, hexagons, or squares. Indeed all three patterns occur in RBC as illustrated in Figure 4.1 [6, 7, 8], albeit under different circumstances. Malkus and Veronis [9] predicted that the stable planform for the case of free boundaries and Boussinesq conditions [10, 11] should be straight rolls rather than, for example, squares or hexagons. The foundation for much of the "modern" work on Rayleigh–Bénard convection was laid during the 1960's by the weakly nonlinear analysis of Schlüter, Lortz, and Busse (SLB) [38] for rigid boundaries, that predicted stable straight rolls above onset also for this realistic case. This prediction is in agreement with experiment, as illustrated by the middle pattern of Figure 4.1 as well as by numerous other experiments. SLB also established that the bifurcation to RBC is supercritical, and gave the initial slope S_1 of the Nusselt number $\mathcal{N} \equiv Qd/\lambda\Delta T = 1 + S_1\epsilon + \mathcal{O}(\epsilon^2)$ (λ is the conductivity of the quiescent fluid and Q is the heat-scurrent density). This was consistent with early measurements, for instance with those of Silveston [13]. Modern measurements such as those shown in Figure 4.2 [14], even within their much greater resolution, are also consistent with a supercritical bifurcation. However, the experimental value of S_1 varies somewhat from one experiment to another and is always somewhat lower than the theoretical prediction (for the data in Figure 4.2 $S_1 = 1.28$ whereas the prediction is $S_1 = 1.43$). Possibly this problem is due to boundary effects at the side wall, but this issue is not entirely settled. Conceptually the next great step forward was the realization by Swift and Hohenberg [15] that the bifurcation to RBC, shown by SLB to be supercritical in the deterministic system, becomes subcritical in the presence of thermal noise. Although at the time the first-order nature of the transition was believed to become significant only within a part per million or so of the transition, thus being out of reach of the experimentalist, good evidence for it has been obtained in very recent experiments [16, 17].

Equally important were seminal experimental contributions during the first five or six decades following Bénard's work. Here I mention only a couple. The heat transport measurements of Schmidt and Milverton [18] confirmed the prediction $R_c = 1708$ with an accuracy of better than 10%. The extensive experiments of Silveston [13] already mentioned above provided data for $\mathcal{N}$ from below onset to $R \simeq 5 \times 10^6$. Silveston also visualized the convection patterns in his apparatus, using the shadowgraph method that has become so very important in more recent times [19, 20, 21]. For additional historical notes, the reader may wish to consult Chapter 3 in this volume, and the informative book by Chandrasekhar [1].

During the last three decades, Rayleigh–Bénard convection (RBC) has become a paradigm for the study of pattern formation [22, 23]. It reveals numerous interesting phenomena in various ranges of $\epsilon \equiv \Delta T/\Delta T_c - 1$. Many of these phenomena have been studied in detail recently, using primarily compressed gases as the fluid, sensitive shadowgraph flow-visualization, digital image analysis, and quantitative heat-flux measurements [20, 24]. I briefly mention some of them here, and then discuss a few of these in greater detail in separate sections below.

Even below onset, thermally driven fluctuations of the temperature and velocity fields about their pure conduction averages provide a fascinating example of critical phenomena in a nonequilibrium system. Twenty-eight years ago, it was already predicted by Swift and Hohenberg [15] that these fluctuations should alter the nature of the bifurcation to RBC, making it subcritical and thus analogous to a first-order phase transition in equilibrium systems. Very recent measurements [16, 17] suggest that this is indeed the case.

Above but close to onset the pattern for a Boussinesq system consists of straight rolls (see Figure 4.1 middle), possibly with some defects induced by the side walls [25]. When non-Boussinesq conditions prevail, a pattern of perfect, defect-free hexagons evolves (see Figure 4.1 left).

Further above onset, for $\epsilon > 0.5$ or so, an interesting qualitatively different state of spatio–temporal chaos, known as spiral–defect chaos (SDC), occurs in systems with Prandtl numbers $\sigma \equiv \nu/\kappa$ of order one or less [26]. This state is a bulk property and not sidewall-induced; it has been studied intensely by theorists as well as experimentalists.

Similarly, RBC was used to study the onset of time dependence over a wide range of σ. [27, 28] Temporally periodic or chaotic patterns were found for $\epsilon > \mathcal{O}(1)$, with the onset occurring at smaller ϵ for smaller σ. However, quantitative studies such as those carried out for SDC are still lacking at larger σ.

The system becomes more complex and interesting even near onset when it is rotated about a vertical axis with an angular velocity Ω. For that case it was predicted [29, 30, 31] and found experimentally [32, 33, 34] that, for $\Omega > \Omega_c$, the primary bifurcation from the conduction state remains supercritical and leads to parallel rolls that are unstable. The instability is to plane-wave perturbations with a wavedirector angle that is advanced relative to that of the rolls by an angular increment Θ_{KL} in the direction of Ω. This phenomenon is known as the Küppers–Lortz (KL) instability. The pattern consists of domains of rolls that incessantly replace each other, both by irregular domain-wall motion and by the KL mechanism. The spatial and temporal behavior suggests the term "domain chaos" for this state. Because this example of spatio–temporal chaos occurs directly at onset, it should be more accessible to theoretical elucidation than, for example, the spiral–defect chaos mentioned above.

Theoretically, the KL instability is expected to persist near onset up to large values of Ω. Thus it was a surprise that the patterns found in experiments near onset changed dramatically when Ω was increased [35]. For $\Omega \geq 70$, there was no evidence of the characteristic domain chaos until ϵ was increased well above

0.1. At smaller ϵ, fourfold coordinated cellular patterns, and in some parameter ranges, slowly rotating, aesthetically appealing, square lattices were encountered.

Relatively unexplored are experimental opportunities that RBC has to offer in the range of σ well below unity. Pure fluids (with rare exceptions [36]) have $\sigma \geq 0.7$. Recently it was shown [37, 38] that smaller values of σ can be reached by mixing two gases, one with a large and the other with a small atomic or molecular weight. The most extreme example readily available is a mixture of H_2 and Xe. Prandtl numbers as small as 0.16 can be reached. In the range $\sigma \leq 0.6$, several interesting new phenomena are predicted to occur [31, 39, 40]. In the $\sigma - \Omega$ plane they include subcritical bifurcations below a line of tricritical bifurcations, Hopf bifurcations to standing waves, a line of codimension-two points where the Hopf bifurcation meets the stationary bifurcation, and a codimension-three point where the codimension-two line and the tricritical line meet.

Another rich and interesting modification of the Rayleigh–Bénard system is achieved by inclining the layer relative to gravity [41, 42, 24]. This adds the tilt angle γ as an additional parameter. In this case the basic state consists of heat conduction and a parallel shear flow that breaks the rotational invariance of the usual RBC. Depending on γ and σ, longitudinal, oblique, transverse, and traveling transverse rolls are the possible flow structures at onset.

No doubt I neglected to mention additional important topics associated with RBC. Nonetheless, at this point we proceed to a somewhat more detailed review of a few of the phenomana listed above that I have found particularly interesting.

4.2 Fluctuations near the Onset of Convection

In the usual deterministic description of RBC, based on the Boussinesq or Navier–Stokes equations, all velocities vanish below the onset of convection and the temperature is given by the pure conduction profile. However, the Brownian motion of the atoms or molecules that occurs because the system is at a finite temperature leads to fluctuations of the temperature and velocity fields that have zero mean but finite mean square. When the fluctuation amplitudes are small enough, their interactions with each other can be neglected and the amplitudes can be described well by stochastic linearized hydrodynamic equations [43]. To my knowledge, the first spatially extended nonequilibrium system for which quantitative measurements of these fluctuations were made was electroconvection in a nematic liquid crystal [44]. Soon thereafter, thermally driven fluctuations were observed also for RBC [45] and quantitative measurements of their amplitudes were made [46, 47]. In part these measurements were made possible by the development of experimental techniques for the study of RBC in compressed gases [25, 20]. There it is possible to use sample spacings an order of magnitude smaller than for conventional liquids and kinematic viscosities are relatively small, thus making the systems more susceptible to noise. In addition, maximizing the sensitivity of the shadowgraph method and careful digital image analysis have enhanced the experimental resolution [20].

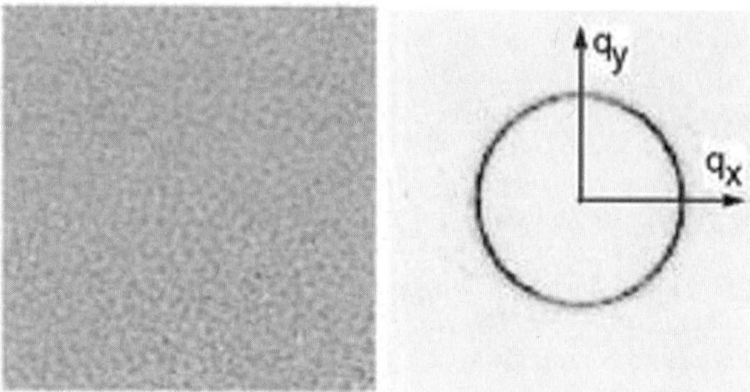

Fig. 4.3. Left: Shadowgraph snapshot of fluctuations below the onset of convection ($\epsilon = -3 \times 10^{-4}$). Right: The average of the square of the modulus of the Fourier transform of 64 images like that on the left. After [46].

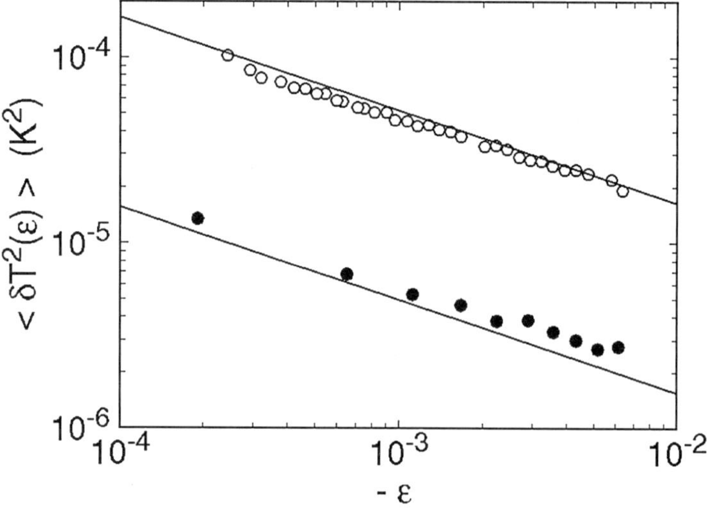

Fig. 4.4. Mean square amplitudes of the temperature fluctuations below the onset of convection of a layer of CO_2 of thickness 0.47 mm and a mean temperature of $32°C$. The solid (open) circles are for a sample pressure of 42.3 (29.0) bars. The two lines are the theoretical predictions. Note that there are no adjustable parameters. After [46].

In the left part of Figure 4.3, we show a processed image of a layer of CO_2 of thickness 0.47 mm at a pressure of 29 bars and at a mean temperature of 32.0°C. The sample was at $\epsilon = -3 \times 10^{-4}$, very close to but just below the bifurcation point. The fluctuating pattern is barely detectable by eye. The right half of the figure shows the average of the structure factors (squares of the moduli of the Fourier transforms) of 64 such images. It demonstrates clearly that the fluctuations have a characteristic wavenumber q. The value of q is in quantitative agreement with the critical wavenumber $q_c = 3.117$ for RBC. The ring in Fourier space is azimuthally uniform, reflecting the continuous rotational symmetry of the RBC system.

The power contained within the ring in Fourier space can be converted quantitatively to the mean-square amplitude of the temperature field [46, 20, 24]. Results for the temporal and spatial averages $\langle \delta T^2 \rangle$ of the square of the deviations of the temperature from the local time average (pure conduction) as a function of ϵ at two different sample pressures are shown in Figure 4.4 using logarithmic scales. The data can be described quite accurately by straight lines with slopes close to $-1/2$, consistent with the powerlaw $\langle \delta T^2 \rangle \propto \epsilon^{-1/2}$ as predicted by linear theory.

The amplitudes of the fluctuating modes below but close to the onset of RBC were calculated quantitatively from the linearized stochastic hydrodynamic equations [43] by van Beijeren and Cohen [48], using realistic (no-slip) boundary conditions at the top and bottom of the cell. For the mean square temperature fluctuations their results give [49, 46]

$$\langle \delta T^2(\epsilon) \rangle = \tilde{c}^2 \left(\frac{\Delta T_c}{R_c} \right)^2 \frac{F}{4\sqrt{-\epsilon}} \,, \tag{4.2}$$

with $\tilde{c} = 3q_c\sqrt{R_c} = 385.28$. Here $R_c = 1708$ is the critical Rayleigh number, and the noise intensity F is given by

$$F = \frac{k_B T}{\rho d\nu^2} \times \frac{2\sigma q_c}{\xi_o \tau_o R_c} \,, \tag{4.3}$$

with $\xi_o = 0.385$ and $\tau_o \simeq 0.0796$. One sees that F depends on the density ρ and kinematic viscosity ν, as well as on the Prandtl number $\sigma = \nu/D_T$ (D_T is the thermal diffusivity). Using the fluid properties of the experimental samples [20], one obtains the straight lines in Figure 4.4. Because there are no adjustable parameters, the agreement between theory and experiment can be regarded as excellent. This agreement lends strong support to the validity of Landau's stochastic hydrodynamic equations [43].

Sufficiently close to the bifurcation, where fluctuation amplitudes become large, nonlinear interactions between them play a role and linear theory breaks down. In this regime genuine critical phenomena that differ from the linear predictions are expected, and the precise critical behavior should depend on the symmetry properties and the dimensionality of the system. Deviations from the prediction of linear theory have been observed recently for electroconvection in nematic liquid crystals [50, 51] that is exceptionally susceptible to the influence of thermal noise. Unfortunately, to this day there are no predictions of the critical phenomena to be expected for this interesting group of systems. For RBC, Swift and Hohenberg were able to show that the system belongs to the same universality class as one considered by Brazovskii [15, 49, 52]. Equilibrium systems belonging to this class include the crystallization of diblock copolymers [53]. For this universality class the transition is of second order at the mean-field level, but the fluctuations induce a first-order tarnsition. A common feature of all the systems belonging to this class is that the order parameter near the bifurcation has a relatively large volume of phase space accessible to it. In the RBC case this is reflected in the rotational invariance of the system as demonstrated by the ring in Fourier space shown in Figure 4.3. On the basis of this qualitative consideration one would not expect the electroconvection system mentioned above [50, 51] to belong to the Brazovskii universality class because the anisotropy due to the director leads to only one or two pairs of spots in Fourier space.

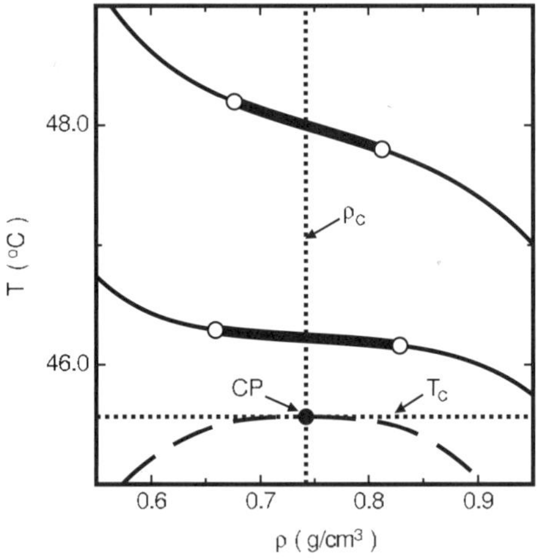

Fig. 4.5. The temperature-density plane near the critical point of SF_6. The dashed line is the coexistence curve separating liquid and vapor. The vertical dotted line is the critical isochore. The solid circle is the critical point $T_c = 45.567°$C, $P_c = 37.545$ bars, and $\rho_c = 0.742$ g/cm^3. The solid lines represent the isobars $P = 38.10$ bars (lower line) and 39.58 bars (upper line) used extensively in experiments. The heavy solid lines, each ending in two circles, illustrate the density range spanned during measurements with $\Delta T \simeq \Delta T_c$ for a cell of spacing $d = 34.3\,\mu$m (lower line) and $d = 59.1\,\mu$m (upper line).

For RBC in ordinary liquids one can estimate [15] that nonlinear fluctuation effects should be observable typically only for $|\epsilon| \leq 10^{-6}$, that has not been accessible to experiments so far. For RBC in compressed gases the critical region is a bit wider, reaching as far out as $|\epsilon| \simeq 10^{-5}$; but as can be seen from Figure 4.4, this too has been beyond experimental resolution. However, the situation is much more favorable near a liquid–gas critical point (CP) [16, 17]. Part of the reason for this can be seen by inspecting Equation 4.3. and the phase diagram of SF_6

shown in Figure 4.5. In that figure we see the temperature-density plane near the CP. The vertical dotted line corresponds to the critical isochore, and the two solid lines are isobars. As the CP is approached on the critical isochore from higher temperatures, the viscosity ν has only a mild singularity and remains finite, whereas the Prandtl number $\sigma = \nu/D_T$ diverges because the thermal diffusivity D_T vanishes. Thus the divergence of σ at finite ν leads to a divergence of F [54]. An equally important aspect is, however, that the fluid properties are such that typical sample spacings d that can be used are in the range of 10 to 100 μm, thus increasing F by one or two orders of magnitude compared to liquids and compressed gases away from the critical point. Another factor that greatly increases the experimental shadowgraph resolution near the CP is the value of the temperature derivative of the refractive index dn/dT. Typically we have $|dn/dT| \simeq 0.1$, whereas for ordinary fluids it tends to be two or three orders of magnitude smaller.

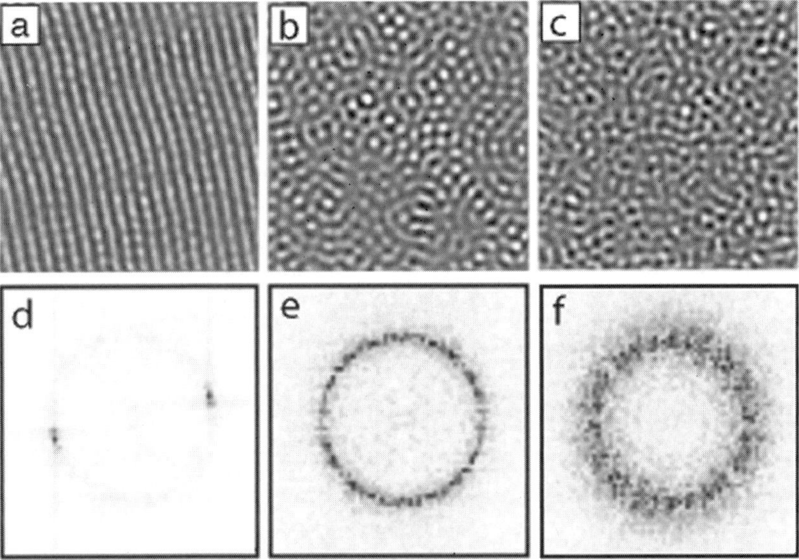

Fig. 4.6. Shadowgraph images (top row) of a 1.28×1.28 mm^2 part of a sample with $d = 34.3$ μm, and the moduli of their Fourier transforms (bottom row). From left to right, the images are for $\epsilon = 0.008, -0.001, and -0.047$. The mean temperature and the pressure corresponded to the critical isochore at $T = 46.22°$C. Adapted from [16].

In Figure 4.6 we show shadowgraph snapshots of fluctuations and roll patterns for a cell of spacing $d = 34.3$ μm at a pressure $P = 38.10$ bars corresponding to the lower isobar shown in Figure 4.5 [16]. The mean temperature $\bar{T} = 46.22°$C was kept constant during the experiment and had a value that corresponded to the critical isochore. When the applied temperature difference was equal to $\Delta T_c = 0.131°$C, the sample occupied the heavy section of the line representing the isobar. The theoretical value of F was 5×10^{-4} for this case.

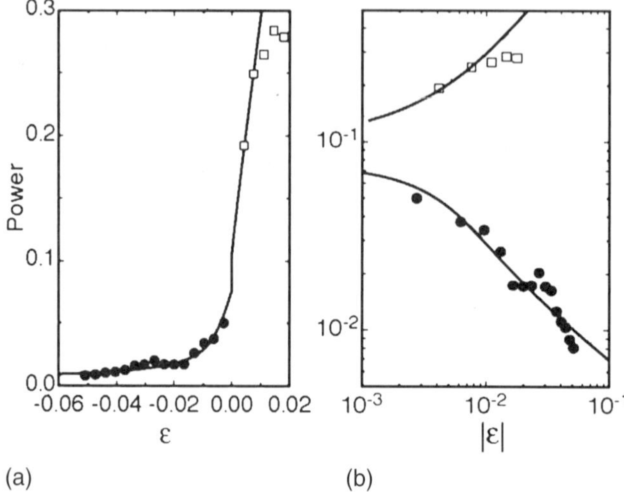

(a)   (b)

Fig. 4.7. Shadowgraph power $\mathcal{P}$ as a function of $\epsilon \equiv \Delta T/\Delta T_c - 1$ for the experiment of Figure 4.6 on (a) linear and (b) logarithmic scales. Solid lines: fit of the Swift–Hohenberg prediction [49] to the data. From [16].

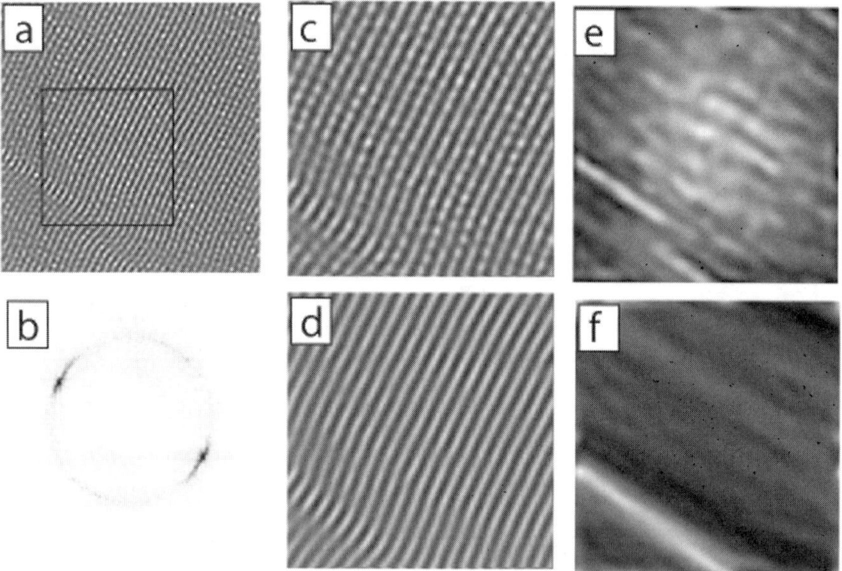

Fig. 4.8. Patterns from a sample with $d = 59\ \mu$m at $\epsilon = 0.009$. (a) Image of size $1.92 \times 1.92\ \mathrm{mm}^2$ and (b) the modulus of its Fourier transform. (c) The $0.96 \times 0.96\ \mathrm{mm}^2$ area inside the square in (a). (d) Same area as (c), but after a bandpass filter was applied around the Fourier-transform peaks of (b). (e) Amplitude of the rolls of (a) obtained by Fourier-transform demodulation. (f) Director angle of (a). Adapted from [16].

The images are for several ϵ values. The bottom row shows the moduli of their Fourier transforms. Just above onset the pattern consisted of convection rolls, as predicted for the deterministic system [38]. Consistent with the Swift–Hohenberg prediction of a first-order transition, there was a sharp transition from a rolls pattern to one of disordered fluctuating cellular structures as ΔT was decreased below ΔT_c.

Figure 4.7 gives results for the shadowgraph power (the square of the modulus of the Fourier transform) as a function of ϵ. One sees a dramatic change in the power at $\epsilon = 0$. The solid lines are a fit of the prediction of Swift and Hohenberg to the data. This fit yielded $F = 7 \times 10^{-4}$, in good agreement with the prediction based on the fluid properties.

Aside from the order of the transition, an issue of considerable interest is the nature of the ordered state (i.e., the rolls) above onset. In Figure 4.8 we show an example [16]. One sees that the rolls reveal several types of disorder. Particularly in the enlarged image Figure 4.8c it can be seen that the rolls were modulated along their axis. This was the result of the superposition of fluctuations of random orientation. As seen in Figure 4.8d, it could be removed by bandpass Fourier filtering with the filters centered on the two peaks of the transform shown in Figure 4.8b. A second type of disorder took the form of an amplitude modulation that varied irregularly in time and space. A snapshot of the roll amplitude, obtained by Fourier demodulation, is shown in Figure 4.8e. A third type of disorder took the form of roll undulations; that is, a variation of the angle of the roll director along the roll axis. A gray-scale rendering of the director angle, obtained from a local wavedirector analysis [57], is shown in Figure 4.8f. We see that both the roll amplitude and the director-angle modulation are correlated over relatively long distances in the direction of the wavedirector, and vary much more rapidly along the roll axis. Some of this noise-induced disorder had been anticipated by Toner and Nelson [58], and should have a commonality with disorder near phase transitions in other two-dimensional systems.

Interestingly, rolls are encountered above the bifurcation only when the mean sample temperature is such that the density corresponds to the critical density. Figure 4.9 shows the fluctuations just below (left image) and the ordered pattern just above (right image) the bifurcation for an experiment in which the mean temperature was 48.3°C [6]. At the pressure of the experiment (39.58 bars) the critical density would have been achieved at 48.0°C. One sees that a dislocation-free lattice of hexagons forms. Although the hexagons are reminiscent of Henri Bénard's beautiful patterns, they have their origin in non-Boussinesq effects [59, 60] whereas Bénard's hexagons were caused by a temperature-dependent surface tension. Measurements of the hysteresis associated with the formation and disappearance of the hexagons in Figure 4.9, as well as a transition to rolls at larger ϵ, were in quite good agreement with predictions based on the deterministic equations of motion [59] even though fluctuations were present [61].

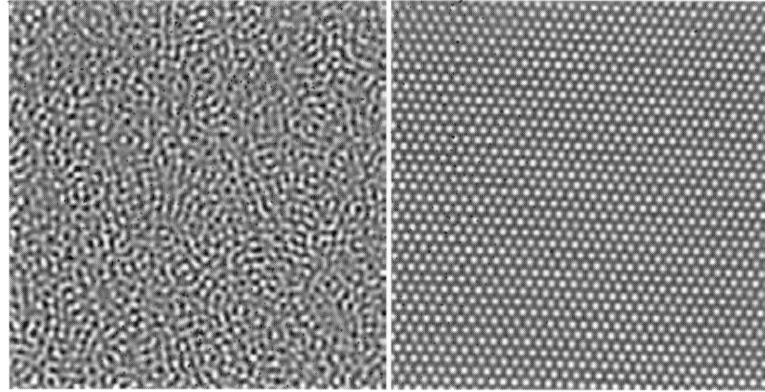

Fig. 4.9. Shadowgraph images of a 1.92×1.92 mm^2 part of a sample with $d = 59$ μm and a pressure of 39.58 bars. Left: $\epsilon = -0.0015$. Right: $\epsilon = 0.0025$. The mean temperature was 48.3°C.

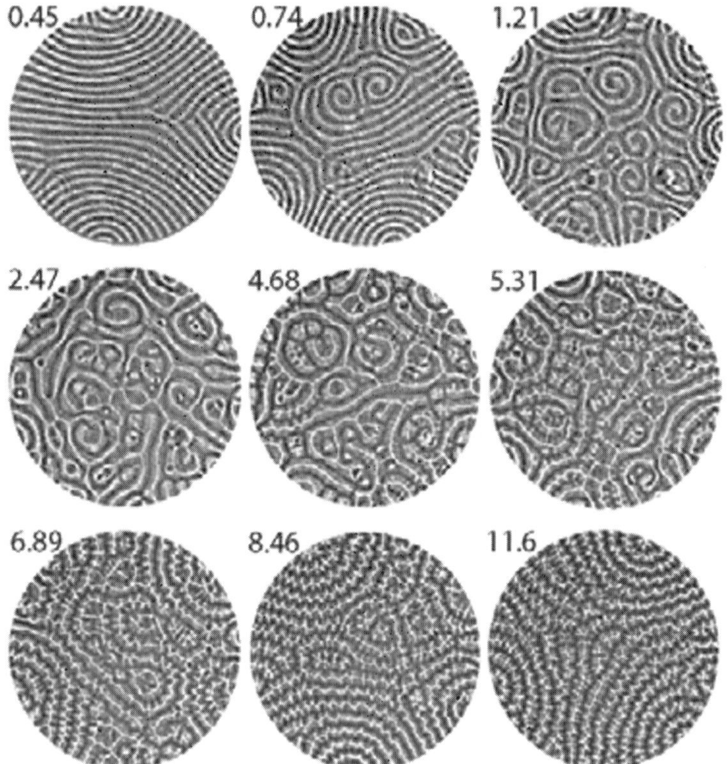

Fig. 4.10. Shadowgraph images for CO_2 at a pressure of 33.25 bars and mean temperature of 21.16°C in a cell with $d = 1.5$ mm and aspect ratio $\Gamma = 28.7$. The Prandtl number was 1.0 and ΔT_c was found to be 0.317°C. The number near each image gives the value of $\epsilon = \Delta T / \Delta T_c - 1$. After [62].

4.3 Deterministic Patterns

When the effective noise intensity is relatively small, the system above onset can be understood in terms of the deterministic equations of motion. The formation of deterministic patterns takes many forms and depends on such parameters as the Prandtl number, the aspect ratio, and the shape of the side walls. Any attempt at a thorough review is well beyond the scope of this chapter. As an example of the richness of pattern-formation phenomena that are encountered, I show in Figure 4.10 some shadowgraph images for $\sigma = 1.0$ and $\Gamma = 28.7$ in a cylindrical cell [62]. For this case $F = 1 \times 10^{-7}$, and stochastic effects do not play an important role. The patterns were obtained with compressed CO_2 as the fluid, but the values of σ and Γ are fairly close to those studied by Croquette and coworkers [25] using argon under pressure and to those of Hu et al. [63] using CO_2. Some of Croquette's results are shown in the chapter by Manneville in this volume (Chapter 3). Croquette found that a time-independent pattern existed only close to onset, roughly for $\epsilon < 0.12$. As ϵ increased, the rolls developed an increasing tendency to terminate with their axes orthogonal to the side wall. The consequent roll curvature and the associated mean flow caused a compression of the rolls near the cell center. For ϵ close to 0.12 the wavenumber in the interior crossed the skewed-varicose instability boundary [64] and a temporal succession of dislocation pairs was formed, thus rendering the pattern time dependent. Most likely this process provides the explanation of the time dependence observed close to onset by heat-transport measurements in early cryogenic convection experiments [65, 66].

As ϵ increased, the patterns became more complex as illustrated in Figure 4.10 for $\epsilon = 0.45$. Typically three wall foci existed at this point. Because of the associated roll curvature there were mean-flow fields emanating from the foci. These flows were strong enough to cause a continuous emission of traveling convection rolls from the foci, leading to a complicated dynamics in the cell interior [63]. These patterns were, however, sidewall-induced and not intrinsic to the interior of a very large system. This was shown in an experiment where the walls were replaced by a very gentle radial ramp in the cell spacing that led to a region of pure conduction surrounding the convecting interior [7]. An example of a pattern in this system, for $\epsilon = 0.21$, is shown in the middle of Figure 4.1. In that case one found time-independent near-perfect rolls without defects and with relatively little roll curvature.

Somewhere near $\epsilon = 0.8$ a new phenomenon occurred. Small spirals formed in the interior, as illustrated in Figure 4.10 for $\epsilon = 0.74$ and 1.21. The formation of these spirals was an intrinsic property of the bulk convection system and was not induced by the side walls. This state, known as spiral–defect chaos, has been known to exist only for the last decade or so [26] and is discussed in more detail in Section 4.4.2.

As ϵ increased further, the structures became more disordered and the spirals were a less dominant feature as seen at $\epsilon = 2.47$. The next interesting phenomenon was first noticeable for $\epsilon = 4.68$, and became more pronounced as ϵ increased to the larger values. This was a transverse perturbation of the convection rolls by a moduation that had a relatively short wavelength. This new feature was due to the oscillatory instability predicted by Clever and Busse [67]. These transverse modulations of the rolls were traveling waves that moved along the roll axes.

The evolution with increasing ϵ seen for the last three patterns is remarkable. Although the patterns became more complex in the sense that the oscillatory modulation became more pronounced, on a coarse-grained scale that averages over the traveling waves they became simpler again. Thus, the pattern at $\epsilon = 11.6$ was not unlike the one for $\epsilon = 0.45$; both had three wall foci and similar defect structures in the interior. It would be nice to be able to understand this reduction of complexity with increasing stress.

4.4 Spatio-Temporal Chaos

4.4.1 Early Measurements

The early 1970s brought a broad survey over a wide range of Prandtl numbers of the occurrence of time-dependent patterns in RBC [27, 28]. At about that time quantitative studies of the statistical properties of spatio–temporal chaos (STC) for σ near one were carried out on RBC at cryogenic temperatures [68, 69, 70, 71]. This early work was followed soon by quantitative measurements [72, 73] on temporal chaos in systems without significant spatial extent that, for some time, attracted far more attention because they made contact with concurrent theoretical developments [74]; this interaction between theory and experiment revived the field of dynamical systems as a branch of physics [75]. By now this field has reached a certain level of maturity. Here I want to examine some of the experimental results on chaos in systems with significant spatial variation. For these the level of theoretical understanding is still much more limited than it is for dynamical systems [76].

Results for the time-averaged Nusselt number $\langle \mathcal{N} \rangle$ during the early cryogenic experiments (for which there was no flow visualization) are shown in Figure 4.11a as a function of $\epsilon \equiv \Delta T / \Delta T_c - 1$. A surprise at the time of those measurements was that the convection depended nonperiodically on the time t already at the relatively small values $\epsilon \simeq 1$. This is illustrated in Figure 4.11b for a circular cell with an aspect ratio Γ (radius/height) $= 5.3$ and $\epsilon = 1.23$. The power spectrum of $\mathcal{N}(t)$ was broad, with a maximum at the frequency $f = 0$, and for large f it fell off as f^{-4} as shown in Figure 4.11c. The experimentally observed algebraic falloff was surprising because simple models of chaos in deterministic systems with relatively few degrees of freedom, such as the Lorenz model, have a spectrum with an exponential falloff [77, 78]. It seems likely [69] that the onset of time-dependence was associated with an adjustment of the wavenumber k as a function of ϵ that caused the system to cross an instability boundary, from our present vantage point most likely the skewed-varicose (SV) instability [64]. The apparently algebraic falloff of the spectrum presumably is then attributable to the presence of a large number of chaotic interacting modes in the spatially

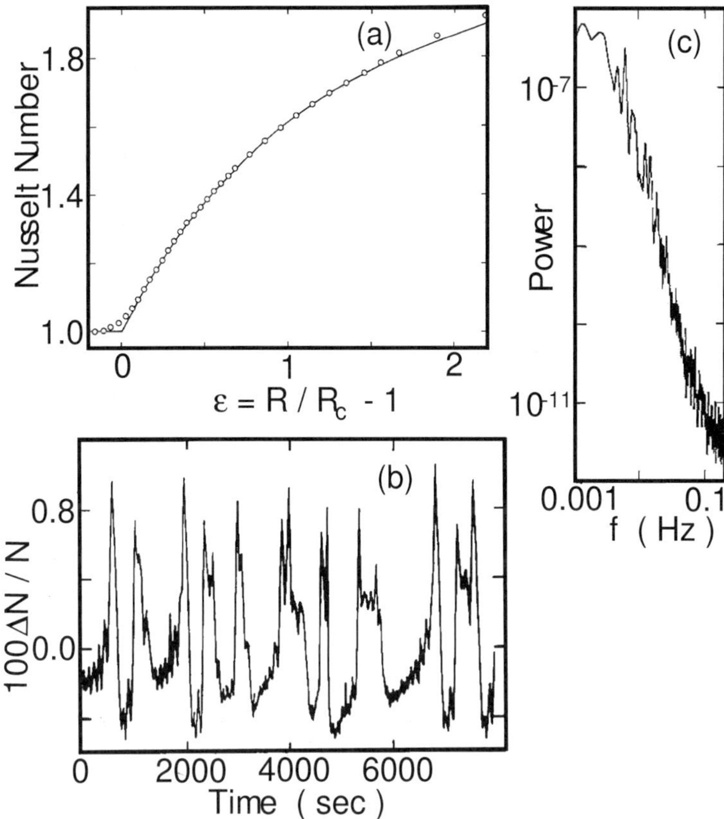

Fig. 4.11. Results from RBC at cryogenic temperatures. (a) The time-averaged Nusselt number as a function of ϵ. (b) Typical deviations of the Nusselt number from its mean for $\epsilon = 1.23$ as a function of time. (c) The power spectrum of a longer sequence of data like those in (b) for $\epsilon = 1.23$. After [70].

extended system that turns out to lead to effectively algebraic decay [77, 78] over the experimentally accessible range of f; but as far as I know a quantitative explanation of this phenomenon is still lacking. In a qualitative sense this suggestion that many modes come into play as the spatial extent increases is an early indicator that spatio–temporal chaos is high-dimensional, and perhaps extensive in the sense that the number of modes (or basis functions) needed to describe it is proportional to (or at least increases with) the system size [79].

In order to provide a quantitative characterization of the chaotic state, the square root of the variance $\sigma_{\mathcal{N}}$ of $\mathcal{N}(t)$ as well as the first moment f_1 of its power spectrum were determined as a funtion of R. As R increased, it turned out that the chaotic state was entered with a discontinuous jump of $\sigma_{\mathcal{N}}$ from zero, and that f_1 was finite at onset. With increasing R, f_1 followed a powerlaw over the two decades $1 < \epsilon < 200$, with an exponent close to $2/3$. To this day I am not aware of a theoretical explanation of these interesting quantitative experimental results. It is also noteworthy that these experiments [68, 69] represent one of the very early examples of computer control of experiments with automated data acquisition [80]. Without this automation it would not have been possible to obtain the results. Similarly, the use for the analysis of experimental results of fast Fourier-transform techniques, that were still relatively new, was a novel feature of this work.

Also still unexplained is the fact that the system remains in the chaotic SV-unstable regime, instead of reducing its wavenumber so as to enter once more a regime of stable rolls that is known to exist for smaller k [64]. This latter phenomenon occurs in the one-dimensional case of a narrow rectangular cell where the SV instability leads to the expulsion of a roll pair and a consequent reduction of the wavenumber. Presumably the is the result of an as yet unknown wavenumber selection process in the two-dimensional system with circular side walls that forces the pattern to remain in the unstable regime. Another feature of the data that was surprising at the time is that the chaos in this system was not preceded by periodic and/or quasi-periodic states that were considered typical of low-dimensional chaotic systems [72]. The absence of these states is consistent, however, with the crossing of an instability boundary that suddenly moves the system into a regime of high-dimensional chaos.

4.4.2 Spiral-Defect Chaos

In spite of its provocative early results and numerous experimental advantages [70, 71], the cryogenic work on STC had its limitations because it did not permit flow visualization. Modern experiments on RBC near ambient temperatures have used the shadowgraph method [19, 20] to visualize the temperature field associated with the convection. Recent experiments on RBC in compressed gases with Prandtl numbers σ close to one led to the discovery [26] that a chaotic state called "spiral–defect chaos" (SDC) is entered at modest ϵ when Γ is large. An example of a shadowgraph image of SDC is shown in Figure 4.12a. SDC consists of many small spirals, targets, and other defects in the roll structure. The defects have a modest lifetime and drift about irregularly, and new ones are constantly created as old ones disappear. The spirals coexist with regions of more or less straight rolls. For the ϵ value of Figure 4.12 a these regions have a width of only a few wavelengths; but near the onset of SDC, and particularly for very large aspect ratiocells [81], the straight-roll regions can become quite large. By now the SDC state has been studied in other experiments that are too numerous to list at this point. A recent review of much of this work and numerous references may be found in [24]. SDC also has been found in numerical solutions of model equations [82, 83] and of the Boussinesq equations [84]. Here I mention only one interesting aspect of this state. Figure 4.12b shows the azimuthal average of the structure factor $S(k)$ (square of the modulus of the Fourier transform) of SDC images. $S(k)$ can be used to compute the mean wavenumber $\bar{k}$. Results for $\bar{k}$ are shown as a function of ϵ in Figure 4.12c. One can see that all the results for $\bar{k}$ lie well within the range where straight rolls are also known to be stable [67, 64]. Thus we arrive at the interesting conclusion that SDC is not caused by a bulk instability of the straight-roll patterns as apparently was the case in the smaller aspect ratio cryogenic experiments. Instead there is bistability of SDC and the usual roll state, that is, over a wide parameter range straight rolls (a fixed point) as well as SDC (a chaotic attractor) are stable solutions of the equations of motion of the system. For Prandtl numbers close to or less than one it turns out that the initial and boundary conditions of typical experiments fall within the attractor basin of SDC, and that rolls without spirals are rarely observed for ϵ greater than some onset value ϵ_s [85].

A quantitative understanding of SDC has not been achieved so far. The problem is very difficult because the chaotic state evolves from a ground state that is already extremely complex (see, e.g., the upper left image of Figure 4.10). However, some insight into the dynamics of this state has been gained. Is seems likely that mean-flow fields play a significant role [86, 24, 87]. A central feature of the dynamics seems to be the competition between two wavenumber selection processes [83]. The spiral tip selects one wavenumber, and the far field that is dominated by a number of different defect types selects another. The resulting wavenumber gradient orthogonal to the spiral arms leads to outward traveling waves surrounding the spiral tips that are equivalent to spiral rotation.

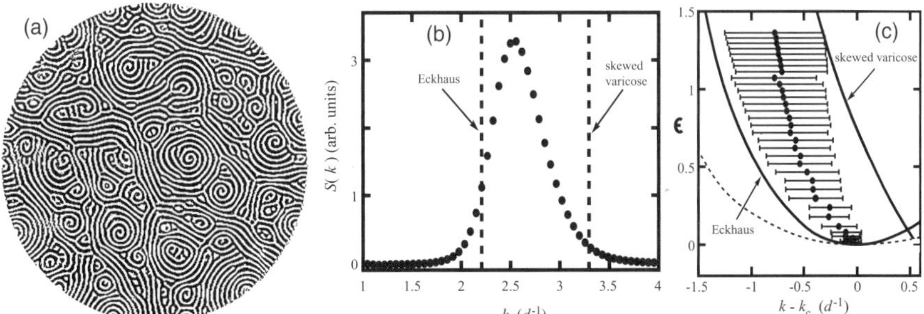

Fig. 4.12. Spiral–defect chaos. (a) Shadowgraph image for $\Gamma = 78$, $\sigma = 0.96$, and $\epsilon = 0.72$. (b) Structure factor $S(k)$ of images like that in (a), but for $\epsilon = 0.46$ (vertical dotted lines are stability boundaries of straight rolls). (c) $\bar{k}$ as a function of ϵ (solid lines are the Eckhaus and skewed-varicose instability of straight rolls; horizontal bars are the widths of $S(k)$). After [26].

4.5 Effect of Rotation

4.5.1 Domain Chaos

As mentioned in the introduction, RBC becomes even more complex and interesting when the sample is rotated about a vertical axis. In that case, the

Coriolis force must be added to the equation of motion (the centrifugal force usually is neglected because to lowest order it is balanced by a pressure gradient sustained by the side wall). The result is that, for $\Omega > \Omega_c$, the rolls that form above onset are unstable to plane-wave perturbations with a wavedirector that has a characteristic angle Θ_{KL} relative to the roll wave director. For $\sigma \geq 0.33$, the bifurcation is expected to be supercritical both below and above Ω_c. Thus the KL instability offers a rare opportunity to study STC in a system where the average flow amplitude evolves continuously from zero and where weakly nonlinear theories are expected to be applicable. After receiving only limited attention for several decades [29, 30, 31, 32, 88, 89, 90], the opportunity to study STC has led to a recent increase in activity both theoretically and experimentally [20, 33, 34, 91, 92, 93, 94, 95, 96, 97]. Indeed, as predicted theoretically [29], the straight rolls at the onset of convection for $\Omega > \Omega_c$ are found to be unstable. In the spatially extended system this leads to the coexistance of domains of rolls of more or less uniform orientation with other domains of a different orientation [32, 88]. A typical example is shown in Figure 4.13b. The replacement of a given domain of rolls proceeded primarily via domain-wall propagation. More recently the KL instability was investigated with shadowgraph flow-visualization very close to onset. It was demonstrated that the bifurcation is indeed supercritical, and that the instability leads to a continuous domain switching through a mechanism of domain-wall propagation also at small ϵ [98, 33, 99, 34]. This qualitative feature has been reproduced by Tu and Cross [93] in numerical solutions of appropriate coupled Ginzburg–Landau (GL) equations, as well as by Neufeld et al. [95] and Cross et al. [96] through numerical integration of a generalized Swift–Hohenberg (SH) equation. There is, however, also a contribution to the dynamics from nucleation of dislocation pairs via the KL mechanism [100].

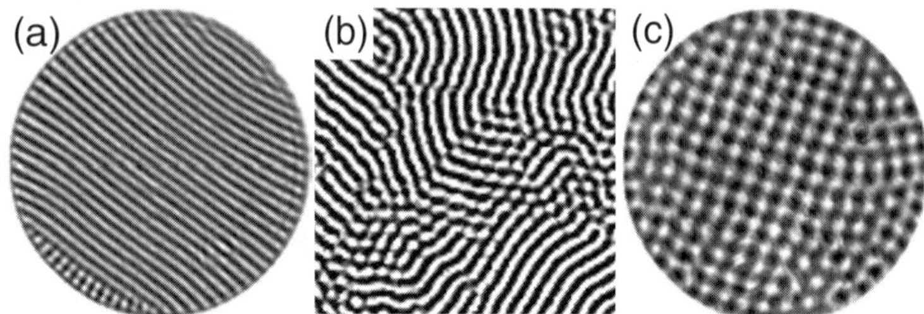

Fig. 4.13. Convection patterns for small ϵ. (a) is for $\Omega = 0$ and Ar gas with $\sigma = 0.69$ and $\epsilon = 0.07$ (from [37]). It shows the predicted [38] straight-roll pattern. (b) is for $\Omega = 15.4$ and CO_2 at a pressure of 32 bar with $\sigma = 1.0$ and $\epsilon = 0.05$ (from [33]). It is a typical pattern in the Küppers–Lortz unstable range. (c) is for argon at 40 bar with $\sigma = 0.7$, $\Omega = 145$, and $\epsilon = 0.04$ (from [35]); it shows no evidence of the Küppers–Lortz instability, and instead consists of a slowly rotating square lattice.

Central features of the KL STC are the time and length scales of the chaotic state near onset. The GL model assumes implicitly a characteristic time-

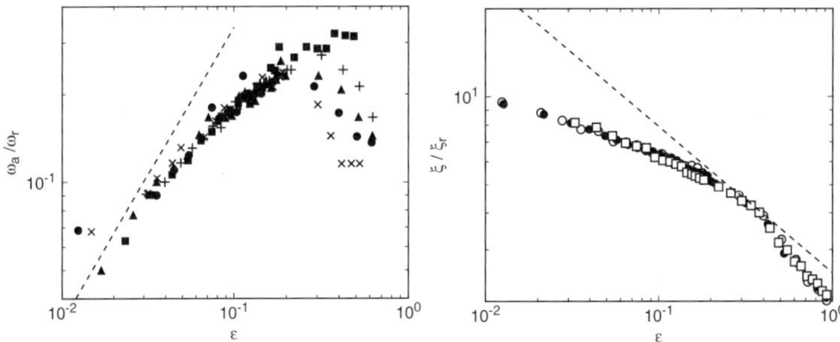

Fig. 4.14. The characteristic frequencies ω_a (left) and lengths ξ (right) of the KL state. The data were divided by Ω-dependent constants ω_r and ξ_r so as to collapse them onto single curves. The dashed lines are shown for reference and have the slopes 1 for ω_a and $-1/2$ for ξ that correspond to the theoretically expected exponents of the time and length scales near onset. The data sets cover approximately the range $14 \leq \Omega \leq 20$. See [33, 99, 34] for details. .

dependence that varies as ϵ^{-1} and a correlation length that varies as $\epsilon^{-1/2}$. Measurements of a correlation length given by the inverse width of the square of the modulus of the Fourier transform as well as a domain-switching frequency as revealed in Fourier space yielded the data in Figure 4.14 [33, 99]. These results seem to be inconsistent with GL equations because they show that the time in the experiment scales approximately as $\epsilon^{-1/2}$ and that the two-point correlation length scales approximately as $\epsilon^{-1/4}$. These results also differ from numerical results based on a generalized SH equation [96] although the range of ϵ in the numerical work is rather limited. We regard the disagreement between experiment and theory as a major problem in our understanding of STC.

4.5.2 Square Patterns at Modest σ

Motivated by the unexpected scaling of length and time with ϵ for the KL state at $\Omega \leq 20$, new investigations were undertaken recently in which the range of Ω was significantly extended to larger values. Contrary to theoretical predictions [31, 39, 101] based on Galerkin procedures and on the stability of appropriate coupled GL equations, it was found [35] that for $\Omega \geq 70$ the nature of the pattern near onset changed qualitatively although the bifurcation remained supercritical. Square patterns like the one shown in Figure 4.13c were stable, instead of typical KL patterns like the one in Figure 4.13b. The squares occurred both when argon with $\sigma = 0.69$ was used and when the fluid was water with $\sigma \simeq 5$. They were observed as well in He-Xe gas mixtures with $\sigma \simeq 0.5$ [102]. For some parameter ranges the lattice was quite disordered; but the fourfold nearest-neighbor coordination remained. The occurrence of squares in this system is completely unexpected and not predicted by theory; the KL instability should continue to be found near onset also at these higher values of Ω. Thus the experiments have uncovered a qualitative disagreement with theoretical predictions in a parameter range where one might have expected the theory to be reliable. Interestingly, very recent direct numerical simulations based on the Boussinesq equations have reproduced the square patterns near onset [24].

A further interesting aspect of the square patterns is that the lattice rotates slowly relative to the rotating frame of the apparatus. This was found in the experiments with argon and water [35] as well as in the simulation [24]. Measurements of the angular rotation rate ω of the lattice for the water experiment are consistent with $\omega(\epsilon)$ vanishing as ϵ goes to zero. Thus the experimental results do not necessarily imply that the bifurcation to squares is a Hopf bifurcation. Quite possibly, as the aspect ratio of the cell diverges, the slope of $\omega(\epsilon)$ vanishes because an infinitely extended lattice cannot rotate. Alternatively, of course, the lattice might become unstable as Γ becomes large. It would be interesting to study the Γ dependence of ω experimentally. To my knowledge there is as yet no theoretical explanation of this rotation.

4.5.3 The Range $0.16 < \sigma < 0.7$

When a RBC system is rotated about a vertical axis, the critical Rayleigh number $R_c(\Omega)$ increases. $R_c(\Omega)$ is predicted to be independent of σ, and experiment [99] and theory [1] for it are in excellent agreement as shown in Figure 4.15a. For $\sigma > 0.33$, the bifurcation is expected to be supercritical and to lead to KL chaos unless Ω is quite large. As discussed above in Section 4.5.2, recent experiments have shown that this is not the case; for $\Omega \geq 70$ square patterns were found that are unrelated to the typical KL domains. For large Ω and $\sigma < 0.68$, the stationary bifurcation is predicted [39] to be preceded by a supercritical Hopf bifurcation; but for $\sigma > 0.33$ experiments have not yet reached values of Ω sufficiently high to encounter time-periodic patterns.

The experimentally accessible range $0.16 \leq \sigma \leq 0.33$ is truly remarkable because of the richness of the bifurcation phenomena that occur there when the system is rotated. For instance, for $\sigma = 0.26$ there is a range from $\Omega \simeq 16$ to 190 over which the bifurcation is predicted to be subcritical. This is shown by the dashed section of the curve in Figure 4.15c. The subcritical range depends on σ. In Figure 4.15b it covers the area below the dashed curve. Thus, the dashed curve is a *line* of tricritical bifurcations. It has a maximum in the $\Omega - \sigma$ plane, terminating in a "tricritical endpoint". An analysis of the bifurcation phenomena that occur near it in terms of Landau equations may turn out to be interesting. One may expect path-renormalization [103] of the classical exponents in the vicinity of the maximum. We are not aware of equivalent phenomena in equilibrium phase transitions, although presumably they exist in as yet unexplored parameter ranges.

At relatively large Ω, the stationary bifurcation (regardless of whether it is super- or sub-critical) is predicted to be preceded by a supercritical Hopf bifurcation that is expected to lead to standing waves of convection rolls [39]. Standing waves are relatively rare; usually a Hopf bifurcation in a spatially extended system leads to traveling waves. An example is shown by the dash-dotted line near

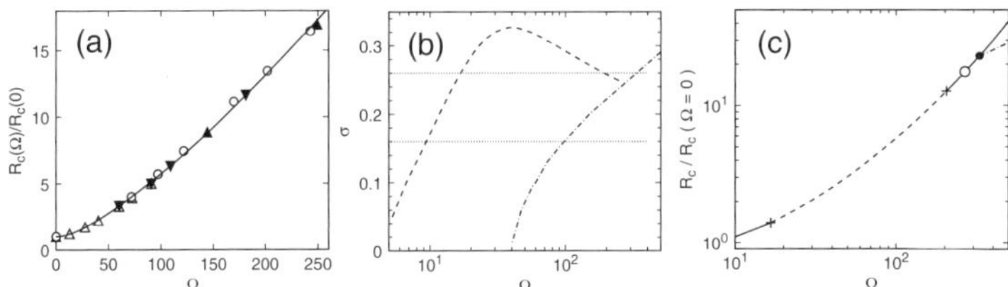

Fig. 4.15. The bifurcation diagram for RBC with rotation about a vertical axis. (a) Experimental and theoretical results for $R_c(\Omega)$ obtained with water (open circles) and Ar at three different pressures (triangles) on linear scales. After [35]. (b) The theoretically predicted bifurcation diagram for RBC with rotation about a vertical axis. The dashed curve gives the tricritical line. The dash-dotted line is the codimension-two line where the Hopf bifurcation meets the stationary bifurcation (e.g., the solid circle in (c)). For $\sigma = 0.24$ the codimension-two line intersects the tricritical line, leading to the codimension-three point shown as an open circle in (c). The upper dotted line in (b) corresponds to the path represented in (c). The lower dotted line in (b) represents the lowest σ-value accessible to experiment using gas mixtures. (c) Bifurcation lines for $\sigma = 0.26$. The dashed line shows the range over which the stationary bifurcation is subcritical. The two plusses are the tricritical points. The dash-dotted line at large Ω shows the Hopf bifurcation. From [40].

the right edge of Figure 4.13b. As can be seen there, the Hopf bifurcation terminates at small Ω at a codimension-two point on the stationary bifurcation that, depending on σ, can be super- or subcritical. The line of codimension-two points is shown in Figure 4.15b as a dash-dotted line. One sees that the tricritical line and the codimension-two line meet at a codimension-three point, located at $\Omega \simeq 270$ and $\sigma \simeq 0.24$. We note that this is well within the parameter range accessible to experiments. We are not aware of any experimentally accessible examples of codimension-three points. This particular case should be accessible to analysis by weakly nonlinear theories, and a theoretical description in terms of GL equations would be extremely interesting and could be compared with experimental measurements.

The σ-range of interest is readily accessible to us by using mixtures of a heavy and a light gas [38]. Values of σ versus the mole fraction x of the heavy component for a typical pressure of 22 bar and at 25°C are shown in Figure 4.4. An important question in this relation is whether the mixtures will behave in the same way as pure fluids with the same σ. We believe that to a good approximation this is the case because the Lewis numbers are of order one. This means that heat diffusion and mass diffusion occur on similar time scales. In that case, the concentration gradient will simply contribute to the buoyancyforce in synchrony with the thermally induced density gradient, and thus the critical Rayleigh number will be reduced. Scaling bifurcation lines by $R_c(\Psi)$ (Ψ is the separation ratio of the mixture) will mostly account for the mixture effect. To a limited extent we showed already that this is the case [37, 38]. In more recent work we have begun to show that the bifurcation line $R_c(\Omega)/R_c(0)$ is indepen-

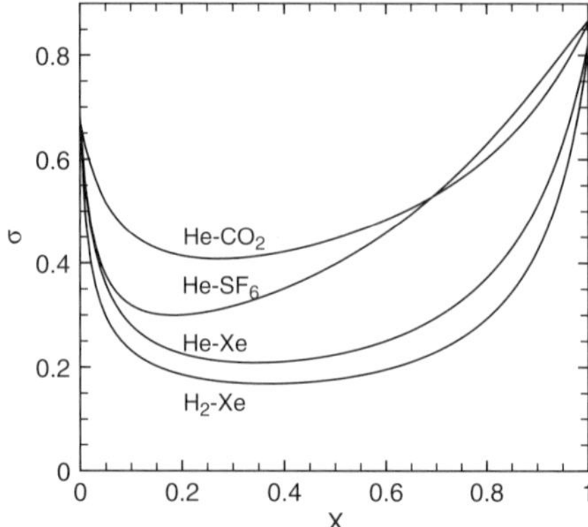

Fig. 4.16. The Prandtl number σ as a function of the mole fraction x of the heavy component for three gas mixtures at a pressure of 22 bar and at 25°C. From [38].

dent of Ψ. Nonetheless we recognize that a theoretical investigation of this issue will be very important.

Assuming that the mixtures behave approximately like pure fluids, we see that the codimension-three point can be reached using either H_2–Xe or He–Xe mixtures. The tricritical point can be reached also using He–SF_6.

4.6 Conclusion

In these few pages, it has been possible to touch only on a few of the interesting aspects of RBC. Some others are discussed in Chapter 3; but even collectively these two contributions do not constitute a thorough review of the field. Nonetheless it is clear that a century of research since the original work of Henri Bénard on this conceptually simple system has strongly advanced our understanding of spatially extended nonlinear dissipative systems. However, much remains to be done. For example, the study of external noise on the system is in its infancy. We believe that the bifurcation to RBC becomes subcritical in the presence of noise, but the influence of noise on the "ordered" state (i.e., the convection rolls) has been examined only qualitatively. It also is apparent that there are a number of unsolved problems. Although we have learned a lot from studies of SDC and domain chaos, the general nature of STC is not understood at a quantitative level. Important issues are whether a description in terms of general principles, perhaps analogous to those of equilibrium statistical mechanics, is on the horizon [104]. We also saw that there are several specific issues on which theory and experiment conflict. These include the characteristic length and time scales of domain chaos and the occurrence at onset of square patterns in the presence of rotation. It will be interesting for future generations of physicists to see what the next century will bring.

References

1. For interesting historical notes on this topics, see S. Chandrasekhar, *Hydrodynamic and Hydromagnetic Stability*, The International Series of Monographs on Physics, Clarendon Press, Oxford (1961).
2. H. Bénard, "Les tourbillons cellularies dans une nappe liquide" *Rev. Gen. Sci. Pure Appl.* **11**, 1261(1900); 1309 (1900); "Les tourbillons cellularies dans une nappe liquide transportant de la chaleur par convection en régime permanent", *Ann. Chim. Phys.* **23**, 62 (1901).
3. For examples, see P. Manneville, "Rayleigh–Bénard convection, thirty years of experimental, theoretical, and modeling work", chapter 3 in this volume, and [1].
4. Lord Rayleigh, "On convective currents in a horizontal layer of fluid when the higher temparture is on the under side", *Phil. Mag.* **32**, 529 (1916).
5. H. Jeffreys, "The stability of a layer of fluid heated from below", *Phil. Mag.* (7) **2**, 833 (1926); "Some cases of instability in fluid motion", *Proc. Roy. Soc. (London)* A **118**, 195 (1928).
6. J. Oh and G. Ahlers, unpublished.
7. K.M.S. Bajaj, N. Mukolobwiez, N. Currier, and G. Ahlers, "Wavenumber Selection and Large-Scale-Flow Effects due to a Radial Ramp of the Spacing in Rayleigh–Bénard Convection", *Phys. Rev. Lett.* **83**, 5282 (1999).
8. M.A. Dominguez-Lerma, G. Ahlers, and D.S. Cannell, "Rayleigh–Bénard convection in binary mixtures with separation ratios near zero", *Phys. Rev. E.* **52**, 6159 (1995).
9. W.V.R. Malkus and G. Veronis, "Finite amplitude cellular convection", *J. Fluid Mech.* **38**, 227 (1958).
10. J. Boussinesq, *Théorie analytique de la chaleur, mise en harmonie avec la thermodynamique et avec la théorie mécanique de la lumiére*, vol.2. (Gauthier-Villars, Paris, 1903).
11. A. Oberbeck, "Über die Wärmeleitung der Flüssigkeiten bei Berüksichtigung der Strömungen infloge von Temperatur Differenzen", *Ann. Phys. Chem.* **7**, 271 (1879).
12. A. Schlüter, D. Lortz, and F. Busse, "On the stability of steady finite amplitude convection", *J. Fluid Mech.* **23**, 129(1965).
13. P. L. Silveston, "Wärmedurchgang in waagerechten Flüssigkeitsschichten,", *Forsch. Ing. Wes.* **24**, 29 (1958); 59 (1958).
14. G. Ahlers, unpublished.
15. J.B. Swift and P.C. Hohenberg, "Hydrodynamic fluctuations at the convective instability", *Phys. Rev. A* **15**, 319 (1977).
16. J. Oh and G. Ahlers, "Thermal-noise effect on the transition to Rayleigh–Bénard convection",*Phys. Rev. Lett.* **91**, 094501 (2003).
17. G. Ahlers and J. Oh, "Critical Phenomena near Bifurcations in Nonequilibrium Systems", *Int. J. Mod. Phys. B* **17**, 3899 (2003).
18. R.J. Schmidt and S.W. Milverton, "On the instability of a fluid when heated from below", *Proc. Roy. Soc. (London) A* **152**, 586 (1935).
19. S. Rasenat, H. Hartung, B.L. Winkler, and I. Rehberg, "The shadowgraph method in convection experiments", *Experiments in Fluids* **7**, 412 (1989).
20. J.R. de Bruyn, E. Bodenschatz, S.W. Morris, S. Trainoff, Y. Hu, D.S. Cannell, and G. Ahlers, "Apparatus for the study of Rayleigh–Benard convection in gases under pressure", *Rev. Sci. Instrum.* **67**, 2043 (1996).
21. S.P. Trainoff and D.S. Cannell, "Physical optics treatment of the shadowgraph", *Phys. Fluids* **14**, 1340 (2002).

22. For a recent review, see for instance, M.C. Cross and P.C. Hohenberg, "Pattern-formation outside of equilibrium", *Rev. Mod. Phys.* **65**, 851 (1993).

23. G. Ahlers, "Over two decades of pattern formation, a personal perspective", in *25 Years of Nonequilibrium Statistical Mechanics*, edited by J.J. Brey, J. Marro, J.M. Rubí, and M. San Miguel (Lecture Notes in Physics, Springer, 1995), pp. 91–124.

24. For a recent review of pattern formation in gas convection, see E. Bodenschatz, W. Pesch, and G. Ahlers, "Recent developments in Rayleigh–Benard convection", *Ann. Rev. Fluid Mech.* **32**, 709 (2000).

25. V. Croquette, "Convective pattern dynamics at low prandtl number .2.", *Contemp. Phys.* **30**, 113 (1989); **30**, 153 (1989).

26. S.W. Morris, E. Bodenschatz, D.S. Cannell, and G. Ahlers, "Spiral defect chaos in large aspect ratio Rayleigh–Bénard convection",*Phys. Rev. Lett.* **71**, 2026 (1993).

27. R. Krishnamurty, "On transition to turbulent convection. 2. Transition to time-dependent flow", *J. Fluid Mech.* **42**, 295 (1970); 309 (1970).

28. R. Krishnamurty, "Some further studies on transition to turbulent convection", *J. Fluid Mech.* **60**, 285 (1973).

29. G. Küppers and D. Lortz, "Transition from laminar convection to thermal turbulence in a rotating fluid layer", *J. Fluid Mech.* **35**, 609 (1969).

30. G. Küppers, "Stability of steady finite amplitude convection in a rotating fluid layer", *Phys. Lett.* **32A**, 7 (1970).

31. R.M. Clever and F.H. Busse, "Non-linear properties of convection rolls in a horizontal layer rotating about a vertical axis",*J. Fluid Mech.* **94**, 609 (1979).

32. F.H. Busse and K.E. Heikes, "Convection in a rotating layer—simple case of turbulence", *Science* **208**, 173 (1980); K.E. Heikes and F.H. Busse "Weakly nonlinear turbulence", Ann. N.Y. Acad. Sci. **357**, 28 (1980).

33. Y. Hu, R.E. Ecke, and G. Ahlers, "Time and length scales in rotating Rayleigh–Bénard convection", *Phys. Rev. Lett.* **74**, 5040 (1995); and "Convection under rotation for Prandtl numbers near 1: Linear stability, wave-number selection, and pattern dynamics" *Phys. Rev. E* **55**, 6928 (1997).

34. Y. Hu, W. Pesch, G. Ahlers, and R.E. Ecke, "Convection under rotation for Prandtl numbers near 1: Küppers–Lortz instability",*Phys. Rev.* **58**, 5821 (1998).

35. K.M.S. Bajaj, J. Liu, B. Naberhuis, and G. Ahlers, "Square patterns in Rayleigh–Bénard convection with rotation about a vertical axis", *Phys. Rev. Lett.* **81**, 806 (1998).

36. One exception is liquid helium. As the superfluid-transition temperature 2.176 K is approached from above, σ vanishes. However, experiments are difficult because σ varies from a value of order one to zero over a narrow temperature range of a few mK, and because of the problem of flow visualization, that has only recently been achieved under the required cryogenic conditions [A.L. Woodcraft, P.G.J. Lucas, R.G. Matley, and W.Y.T. Wong, "Visualization of convective flow patterns in liquid helium", *J. Low Temp. Phys.* **114**, 109 (1999)]. Other exceptions are liquid metals that have $\sigma = \mathcal{O}(10^{-2})$ because of the large electronic contribution to the conductivity. However, it is not possible to explore the range $10^{-2} \lesssim \sigma \lesssim 0.7$ with them. Because liquid metals are not transparent to visible light, flow visualization is also a problem.

37. J. Liu and G. Ahlers, " Spiral-defect chaos in Rayleigh–Benard convection with small Prandtl numbers", *Phys. Rev. Lett.* **77**, 3126 (1996).

38. J. Liu and G. Ahlers, "Rayleigh–Bénard convection in binary-gas mixtures: Thermophysical properties and the onset of convection", *Phys. Rev. E* **55**, 6950 (1997).

39. T. Clune and E. Knobloch, "Pattern selection in rotating convection with experimental boundary-conditions", *Phys. Rev. E* **47**, 2536 (1993).

40. K.M.S. Bajaj, G. Ahlers, and W. Pesch, "Rayleigh–Bénard convection with rotation at small Prandtl numbers", *Phys. Rev. E* **65**, 056309 (2002).

41. F.H. Busse and R.M. Clever, "3-dimensional convection in an inclined layer heated from below", *J. Eng. Math.* **26**, 1 (1992).

42. R.E. Kelly, "The onset and development of thermal convection in fully developed shear flows", *Adv. in Appl. Mech.* **31**, 35 (1994).

43. L.D. Landau and E.M. Lifshitz, *Mekhanika Sploshnykh Sred (Fluid Mechanics)* Translated into English Oxford: Pergamon Press, (1959).

44. I. Rehberg, S. Rasenat, M. de la Torre-Juarez, W. Schöpf, F. Hörner, G. Ahlers, and H.R. Brand, "Thermally induced hydrodynamic fluctuations below the onset of electroconvection", *Phys. Rev. Lett.* **67**, 596 (1991).

45. E. Bodenschatz, S. Morris, J. de Bruyn, D.S. Cannell, and G. Ahlers, "Convection in gases at elevated pressures", in *Pattern Formation in Complex Dissipative Systems, Fluid Patterns, Liquid Crystals, Chemical Reactions*, edited by S. Kai (World Scientific, Singapore, 1992), pp. 227–237.

46. M. Wu, G. Ahlers, and D.S. Cannell, "Thermally-induced fluctuations below the onset of Rayleigh–Bénard convection", *Phys. Rev. Lett.* **75**, 1743 (1995).

47. Recent measurements by W. Schöpf and I. Rehberg ["The influence of thermal noise on the onset of traveling-wave convection in binary-fluid mixtures - an experimental investigation", *J. Fluid Mech.* **271**, 235 (1994)] using binary mixtures and by G. Quentin and I. Rehberg ["Direct measurement of hydrodynamic fluctuations in a binary mixture", *Phys. Rev. Lett.* **74**, 1578 (1995)] involved pseudo-one-dimensional sample geometries for which theoretical predictions are more difficult to obtain due to the influence of the side walls. In this review we consider only results obtained for systems of large extent in two dimensions.

48. H. van Beijeren and E.G.D. Cohen, "The effects of thermal noise in a Rayleigh–Bénard cell near its first convective instability", *J. Stat. Phys.* **53**, 77 (1988).

49. P.C. Hohenberg and J. Swift, "Effects of additive noise at the onset of Rayleigh–Bénard convection", *Phys. Rev. A* **46**, 4773 (1992).

50. M.A. Scherer, G. Ahlers, F. Hörner, and I. Rehberg, "Deviations from linear theory for fluctuations below the supercritical primary bifurcation to electroconvection", *Phys. Rev. Lett.* **85**, 3754 (2000).

51. M.A. Scherer and G. Ahlers, "Temporal and spatial properties of fluctuations below a supercritical primary bifurcation to traveling oblique-roll electroconvection", *Phys. Rev. E* **65**, 051101 (2002).

52. S.A. Brazovskii, "Phase transition of an isotropic system to a nonuniform state", *Sov. Phys. JETP* **41**, 85 (1975).

53. F.S. Bates, J.H. Rosedale, G.H. Fredrickson, and C.J. Glinka, "Fluctuation-induced 1st-order transition of an isotropic system to a periodic state", *Phys. Rev. Lett.* **61**, 2229 (1988).

54. This feature was noted also by M. Assenheimer and V. Steinberg, "Critical phenomena employed in hydrodynamic problems: A case study of Rayleigh–Bénard convection", *Europhysics News* **27**, 143 (1996).

55. These results differ from measurements by Roy and Steinberg [56] who found hexagons just above onset for seemingly similar experimental parameters.

56. A. Roy and V. Steinberg, "Reentrant hexagons in non-Boussinesq Rayleigh–Bénard convection: Effect of compressibility", *Phys. Rev. Lett.* **88**, 244503-1 (2002).

57. D.A. Egolf, I.V. Melnikov, and E. Bodenschatz, "Importance of local pattern properties in spiral defect chaos", *Phys. Rev. Lett.* **80**, 3228 (1998).

58. J. Toner and D.R. Nelson, "Smectic, cholesteric, and Rayleigh–Benard order in 2 dimensions", *Phys. Rev. B* **23**, 316 (1981).

59. F. Busse, "Stability of finite amplitude cellular convection and its relation to an extremum principle", *J. Fluid Mech.* **30**, 625 (1967).

60. E. Bodenschatz, J.R. de Bruyn, G. Ahlers, and D.S. Cannell, "Transitions between patterns in thermal-convection", *Phys. Rev. Lett.* **67**, 3078 (1991).

61. On the basis of these results it seems likely that the hexagons reported by A. Roy and V. Steinberg ["Reentrant hexagons in non-Boussinesq Rayleigh–Bénard convection: Effect of compressibility", *Phys. Rev. Lett.* **88**, 244503 (2002)] are caused by non-Boussinesq effects due to a non-critical density in their experiment rather than by compressibility effects as suggested by these authors.

62. J. Liu, K.M.S. Bajaj, and G. Ahlers, unpublished.

63. Y. Hu, R. Ecke, and G. Ahlers, "Convection near threshold for Prandtl numbers near one", *Phys. Rev. E* **48**, 4399 (1993); "Behavior of focus patterns in low Prandtl number convection", *Phys. Rev. Lett.* **72**, 2191 (1994); "Transition to spiral-defect chaos in low Prandtl number convection", *Phys. Rev. Lett.* **74**, 391 (1995); "Convection for Prandtl numbers near 1 - Dynamics of textured patterns", *Phys. Rev. E* **51**, 3263 (1995).

64. F.H. Busse and R.M. Clever, "Instabilities of convection rolls in a fluid of moderate prandtl number", *J. Fluid Mech.* **91**, 319 (1979).

65. G. Ahlers and R.P. Behringer, "Evolution of turbulence from Rayleigh–Bénard instability", *Phys. Rev. Lett.* **40**, 712 (1978).

66. G. Ahlers and R.W. Walden, "Turbulence near onset of convection", *Phys. Rev. Lett.* **44**, 445 (1980).

67. R.M. Clever and F.H. Busse, "Transition to time-dependent convection", *J. Fluid Mech.* **65**, 625 (1974).

68. G. Ahlers, "Convective heat transport between horizontal parallel plates", *Bull. Am. Phys. Soc.* **17**, 59 (1972).

69. G. Ahlers and J.E. Graebner, "Time-dependence in convective heat transport between horizontal parallel plates", *Bull. Am. Phys. Soc.* **17**, 61 (1972).

70. G. Ahlers, "The Rayleigh–Bénard Instability at Helium Temperatures", in *Fluctuations, Instabilities, and Phase Transitions*, edited by T. Riste (Plenum, New York, 1974), p. 181.

71. G. Ahlers, "Low-temperature studies of Rayleigh–Bénard instability and turbulence", *Phys. Rev, Lett.* **33**, 1185 (1974).

72. J. Gollub and H.L. Swinney, "Onset of turbulence in a rotating fluid", *Phys. Rev. Lett.* **35**, 927 (1975).

73. A. Libchaber and J. Maurer, "Une expérience de Rayleigh–Bénard de géométrie réduite: multiplication, accrochage et démultiplication de fréquences", *J. Phys. Colloq. C3* **41**, C3-51 (1978).

74. M. Feigenbaum, "Quantitative universality for a class of nonlinear transformations", *J. Stat. Phys.* **19**, 25 (1978).

75. An overview of dynamical systems and temporal chaos can be found, for instance, in *Order within Chaos*, P. Bergé, Y. Pomeau, and C. Vidal, *Order Within Chaos: Towards a Deterministic Approach to Turbulence* (Wiley, New York, 1986).

76. See, for instance, P.C. Hohenberg and B.I. Shraiman, "Chaotic behavior of an extended system", *Physica D* **37**, 109 (1989); and J.P. Gollub, "Pattern-formation—spirals and chaos", *Nature* **367**, 318 (1994).

77. H.S. Greenside, G. Ahlers, P.C. Hohenberg, and R.W. Walden, "A simple stochastic model for the onset of turbulence in Rayleigh–Bénard convection", *Physica* **5D**, 322 (1982).

78. U. Frisch and R. Morf, "Intermittency in non-linear dynamics and singularities at complex times", *Phys. Rev. A* **23**, 2673 (1981).

79. For a recent discussion of the extensive nature of STC, see D.A. Egolf, I.V. Melnikov, W. Pesch, and R.E. Ecke, "Mechanisms of extensive spatiotemporal chaos in Rayleigh–Bérnard convection", *Nature* **404**, 733 (2000).

80. B.D. Wonsiewicz, A.R. Storm, and J.D. Sieber, "Microcomputer control of apparatus, machinery, and experiments", *Bell Syst. Tech. J.* **57**, 2209 (1978).

81. J. Liu and G. Ahlers, unpublished.

82. H.W. Xi, J.D. Gunton, and J. Vinals, "Spiral defect chaos in a model of Rayleigh–Bénard convection", *Phys. Rev. Lett.* **71**, 2030 (1993).

83. M.C. Cross and Y. Tu, "Defect dynamics for spiral chaos in Rayleigh–Bénard convection", *Phys. Rev. Lett.* **75**, 834 (1995).

84. W. Decker, W. Pesch, and A. Weber, "Spiral defect chaos in Rayleigh–Bénard convection", *Phys. Rev. Lett.* **73**, 648 (1994).

85. See, however, R.V. Cakmur, D.A. Egolf, B.B. Plapp, and E. Bodenschatz, "Bistability and competition of spatiotemporal chaotic and fixed point attractors in Rayleigh–Benard convection", *Phys. Rev. Lett.* **79**, 1853 (1997), where the bistability of straight rolls and SDC was demonstrated experimentally.

86. I.V. Melnikov and W. Pesch, unpublished.

87. K.H. Chiam, M.R, Paul, and M.C. Cross, "Mean flow and spiral defect chaos in Rayleigh–Bénard convection", *Phys. Rev. E* **67**, 056206 (2003).

88. K.E. Heikes and F.H. Busse, "Weakly nonlinear turbulence in a rotating convection layer", *Ann. N.Y. Acad. Sci.* **357**, 28 (1980).

89. K. Buhler and H. Oertel, "Thermal cellular convection in rotating rectangular boxes", *J. Fluid Mech.* **114**, 261 (1982).

90. J.J. Niemela and R.J. Donnelly, "Direct transition to turbulence in rotating Bénard convection", Phys. Rev. Lett. **57**, 2524 (1986).

91. F. Zhong, R. Ecke, and V. Steinberg, "Rotating Rayleigh–Bénard convection - Küppers-Lortz transition", Physica D **51**, 596 (1991).

92. F. Zhong and R. Ecke, "Pattern Dynamics and Heat Transport in Rotating Rayleigh–Bénard Convection", *Chaos* **2**, 163 (1992).

93. Y. Tu and M. Cross, "Chaotic domain-structure in rotating convection", *Phys. Rev. Lett.* **69**, 2515 (1992).

94. M. Fantz, R. Friedrich, M. Bestehorn, and H. Haken, "Pattern Formation in Rotating Bnard Convection", *Physica D* **61**, 147 (1992).

95. M. Neufeld, R. Friedrich, and H. Haken, "Order-parameter equation and model equation for high prandtl number—Rayleigh–Bénard convection in a rotating large aspect ratio system", *Z. Phys. B.* **92**, 243 (1993).

96. M. Cross, D. Meiron, and Y. Tu, "Chaotic domains: A numerical investigation", *Chaos* **4**, 607 (1994).

97. Y. Ponty, T. Passot, and P. Sulem, "Chaos and structures in rotating convection at finite Prandtl number", *Phys. Rev. Lett.* **79**, 71 (1997).

98. E. Bodenschatz, D.S. Cannell, J.R. de Bruyn, R. Ecke, Y. Hu, K. Lerman, and G. Ahlers, "Experiments on 3 systems with nonvariational aspects", *Physica D* **61**, 77 (1992).

99. Y. Hu, R.E. Ecke, and G. Ahlers, "Convection under rotation for Prandtl numbers near 1: Linear stability, wavenumber selection, and pattern dynamics", *Phys. Rev. E* **55**, 6928 (1997).

100. N. Becker and G. Ahlers, unpublished.

101. W. Pesch, private communication.

102. K.M.S. Bajaj and G. Ahlers, unpublished.

103. M.E. Fisher, "Renormalization of critical exponents by hidden variables", *Phys. Rev.* **176**, 257 (1968).
104. See, for instance, D.A. Egolf, "Statistical mechanics: Far from equilibrium", *Science* **296**, 1813 (2002).

5 Rayleigh–Bénard Convection as a Model of a Nonlinear System: A Personal View

Yves Pomeau

Laboratoire de Physique Statistique de l'Ecole Normale Supérieure, associé au CNRS, 24 Rue Lhomond, 75231 Paris Cedex 05, France

Summary. This is a (personal) review of work done on Rayleigh–Bénard instability, seen primarily as a model system generating periodic structures in space as more or less parallel convection rolls. This nonequilibrium crystallography was and still is a rich source of inspiration, as I show.

Over the years Rayleigh–Bénard convection has been a major theme of research in nonequilibrium science, both because it is a great experimental model, particularly in the hands of skilled experimentalists such as Pierre Bergé and Monique Dubois, and because the theory follows amplitude equations with a rather simple structure, at least near the onset of instability. This chapter reports work done mostly with Paul Manneville whose collaboration over the years has been invaluable.

Some could complain that this field is too far from the overimportant "applications", something that is not completely true: with Loulergue and Manneville years ago we invented a visualization device of infrared radiation, surprisingly sensitive [1], based upon the distortions of the surface of a convecting liquid due to the Marangoni effect (the one actually observed by Bénard). Of course this device is not competitive with solid-state devices, but who knows what will happen in the future? The sensitivity of this device is such that it shows well the temperature differences across a living human face.

My research on Rayleigh–Bénard instability started, rather curiously, with the writing of a review article [2]. Therein, I tried to understand the topic a little better, which was not easy. At the time there were two antagonistic schools studying thermal convection. One claimed that the roll pattern should be linked to the geometry of the box. This geometrical influence explained everything, except for the observed wavelength increase as the temperature difference got larger (more on that later). The other school neglected completely the geometry of the box and claimed that everything was explainable in terms of secondary instability of the roll pattern, as a function of the wavelength. Another theme was the possible existence of secondary transitions in the dependence of the heat flux with respect to the temperature difference. Of course, with a large enough spreading of the experimental results it is possible to find all sorts of bents and transitions in experimental curves!

Unfortunately, all this deflected attention from more serious topics, such as the actual transition from stationary to time-dependent convection, something that was looked at by the Saclay group only much later.

The first research on convection patterns (I refer here to the so-called large boxes, forgetting anything related to the now classical dynamical systems with a few degrees of freedom, for which one can consult [3] or [4]) tried to understand the possible instabilities of a system of parallel rolls, following an analysis inspired by the classical derivation of the equations of solid-state mechanics by Cauchy, resting on the symmetries of the system. Paul Manneville and I found the two modes of (space) phase diffusion of a pattern of parallel rolls: diffusion parallel and perpendicular to the rolls. The parallel phase equation describes how parallel rolls tend to have a constant wavelength when slightly displaced from their exactly periodic positions, although perpendicular diffusion describes the relaxation of slightly bent rolls. One interesting idea was that the wavelength of marginal stability for perpendicular diffusion should play a special role: a result of Landau [5] on the elasticity of smectic liquid crystals points in the same direction. This analysis was supported by the magnificent experiments of the Saclay group [6] that showed the validity of the amplitude equation approach, from which a particular limit of the phase equation follows at once. This forbade consideration of the wavelength for which $D_\perp$ vanish to be the observed optimal wavelength (the elastic buckling studied in [7] is special: there is an underlying optimization principle and therefore an optimal wavelength for given supercritical conditions). Therefore the question of what determines in practice the optimal wavelength above threshold was still open, given the full interval of possible wavelength for a pattern in an infinite system. A first answer, a bit unexpected [8], was that this wavelength is severely constrained if the vertical boundaries do not conduct heat. This result rests on the existence of a conserved quantity perpendicular to the roll direction, which could be interpreted as a constant pressure in a mechanical system. This explains in particular that all rolls should have the same wavelength in a steady state. However, this does not explain the observation of an unique wavelength above threshold. The explanation of this required two steps.

We did show first [9] that the wavelength of an axisymmetric structure is the one that cancels $D_\perp$: to remain steady bent rolls have to have no tendency to straighten or bend, which implies that $D_\perp = 0$. This rational approach to the problem was free of any unproved principle of optimization, as the much-used maximization of the heat flux.

This showed too the relevance of geometrical effects: a single supercritical wavenumber could be shown, but for slightly bent rolls only. This led us to look at other geometries, including finite box effects and various distortions to exactly straight rolls.

In this respect, a relatively recent contribution is worth mentionning [10]: let us imagine a set of rolls such that the outer one is forced to follow an ellipse drawn by the boundary of a box. The rolls inside the container draw near the external boundary a set of curves parallel to the ellipse (these are fourth-order algebraic curves), the distance between two consecutive rolls being half a wavelength. The geometry is of course very reminiscent of a problem of geometrical optics: each roll can be seen as a wave surface, distant from the other surface by a wavelength.

It is well known in general that such a geometrical pattern (an idea of Huygens) leads to caustics, that is to the fact that more than one wave surface passes through a given point. For the linear wave equation, that more than one wave surface crosses a point only implies that one must add the contribution of each wave, except near the caustics where diffraction must be taken into account.

The case of nonlinear waves, such as the Rayleigh–Bénard rolls, is obviously different, inasmuch as a linear superposition of a different solution does not yield a new solution. The consequence is a rather curious phenomenon from the point of view of the theory of defects in continuous media: the Huygens cusp is replaced there by the end of a grain boundary. This grain boundary is at the merging of two domains where the rolls have different orientations. The jump of orientation across the grain boundary is a function of the distance along this boundary that vanishes at the cusp of the Huygens construction (that is, at a point of convergence of two caustics in geometrical optics). Thanks to this grain boundary, and as expected, only one roll orientation exists everywhere, except precisely near grain boundaries. There, an interesting nonlinear equation replaces the (linear) diffraction problem solved by the Pearcey integral near the Huygens cusp.

Another research theme was the one of slowly varying parameters [11]: this was in some sense a systematic exploration of the various linear wave problems to see what happens in the nonlinear case. There we tried to build up what could be called a nonlinear WKB theory of the Bénard rolls. I review this question, referring the interested reader to the original publication. One tries to understand what happens when the Rayleigh–Bénard rolls are set up in a medium with properties changing slowly in space. More precisely the properties of the system change little over one wavelength, but can change of order 1 over large distances (which forbids using a regular perturbation scheme). For the Rayleigh–Bénard instability this is an academic problem, but it is the usual situation in structures observed in nonlinear optics where the strength of the pump field changes (usually) continuously from the center to the edge of the beam. The published work is about the one–dimensional case, which I recall first. Then I present a few ideas about the more general case of rolls in a medium with parameters slowly changing both parallel and perpendicular to the rolls. To analyze this kind of situation, one assumes that locally the rolls are given by a solution of the full equation, with the value of the control parameter constant and equal to its local value. Therefore the amplitude and wavelength of the rolls are functions of the local conditions, summarized in the slow dependence with respect to x of the control parameter ϵ. In an unbounded medium with a constant ϵ a continuum of solutions exists, parameterized by the wavelength Λ, their amplitude being a function $A_0(\Lambda, \epsilon; x)$. The argument x in A_0 is to recall that this solution depends "fastly" on the position x, practically with the wavelength Λ, much shorter than the typical range of ϵ:

$$\frac{\Lambda}{\varepsilon}\frac{d\varepsilon}{dx} \ll 1 \qquad (5.1)$$

Assuming ϵ constant, this function A_0 is a solution of the primitive equations of the field (Oberbeck–Boussinesq equations for the Rayleigh–Bénard instability). Following the general ideas of the adiabatic method, one looks for a solution by expansion in the small parameter, the gradient of ϵ, near the solution A_0, exactly valid for ϵ constant only. At first-order one finds a solvability condition coming from the fact that the zeroth-order solution has a free continuous parameter, the wavelength. A small variation of this wavelength keeps the system in the range of possible solutions, that generates a zero mode of the linearized problem. The solvability condition that results from this relates practically the slow variations of the wavelength and of ϵ. This solvability condition reads:

$$D(\epsilon, \Psi_x)\frac{d\Psi_x}{dx} + C(\epsilon, \Psi_x)\frac{d\epsilon}{dx} = 0 \tag{5.2}$$

In Equation (5.2), I introduced Ψ, the spatial phase of the roll system. The derivative $\Psi_x = d\Psi/dx$ is the local wavenumber of the roll system. Both this wavenumber and ϵ are in the argument of the quantities C and D, the two coefficients in the linear first-order Equation (5.2). Notice that $\Psi_x = 2\pi/\Lambda$ replaces the wavelength Λ in the argument of the functions.

Whenever the system is uniform in space, that is, when ϵ is independent of x, Equation (5.2) reduces to the longitudinal part of the phase equation [12] in the steady case. Therefore it is not too difficult, at least in principle, to extend the analysis to the 2-D case, that is, to systems with ϵ changing both parallel and perpendicular to the rolls. The idea is to base this on the extension of the regular phase equation to include both parallel and perpendicular diffusion. It is enough to replace the first term in Equation (5.2) by the full diffusion equation (with parallel and perpendicular diffusion). This steady diffusion equation reads:

$$D_{//}\frac{\partial^2\Psi}{\partial x^2} + D_\perp\frac{\partial^2\Psi}{\partial y^2} = 0$$

x being the local coordinate perpendicular to the roll axis, and y the one parallel to it. To write this phase diffusion term in an intrinsic way, one replaces the derivation operator $\partial/\partial x$ by $\mathbf{n}.\nabla$, where $\mathbf{n} = \nabla\Psi/\sqrt{(\nabla\Psi)^2}$, ∇ being the usual vector of coordinates $\partial/\partial X, \partial/\partial Y$ in an arbitrary (but fixed) set of rectangular coordinates (X, Y). Similarly the operator $\partial/\partial y$ becomes $\nabla - \mathbf{n}(\mathbf{n}.\nabla)$. Whence the adiabatic equation for Ψ in a slowly changing medium:

$$D_{//}(\epsilon, \mathbf{n}.\nabla\Psi)(\mathbf{n}.\nabla)^2\Psi + D_\perp(\epsilon, \mathbf{n}.\nabla\Psi)\left(\nabla - \mathbf{n}(\mathbf{n}.\nabla)\right)^2\Psi$$
$$+ \, C(\epsilon, \mathbf{n}.\nabla\Psi)(\mathbf{n}.\nabla)\epsilon = 0 \tag{5.3}$$

Of course, this assumes that one can uniformly define the phase Ψ, which is not possible anymore if certain types of defects are present in the roll system. Let us notice, however, that the grain boundary of [10] is compatible with a uniform continuous phase, but with a singular gradient.

This brings me to the defects in roll structures. Actually these defects have the same topology as defects in layered liquid crystals, as smectics. In roll systems one may build a dynamical theory of defects far more detailed than for real liquid crystals because the underlying equation, Oberbeck–Boussinesq far from the onset of instability and amplitude equations near threshold, is well known. The vanishing of the diffusion coefficient $D_\perp$ brings a considerable complexity to the analysis, as well as the possibility of large-scale flows. For instance, the distortion of the phase near a dislocation cannot be derived from a linear equation, including in the far field [13]. Another interesting issue is the non adiabatic coupling between the fast phase and the slow variation of the external parameter [14]. The grain boundaries are free of this difficulty [15]. Their analysis brings in an interesting result, namely, that the rolls have a typical healing length much shorter along their axis than across it. This has remarkable consequences: it explains that in experiments without outside forcing of the structure, rolls tend to cross the lateral boundary at right angles [8, 16]. Therefore, the equilibrium structures in large boxes usually have defects inside.

Actually, all these theories concerned near-threshold situations, or at least not too far from threshold. A first step to go away from this is to look at discontinuous transitions, so common in real shear flows, for instance. Besides numerical simulations, one possible way of approaching this type of transition theoretically is to take inspiration from simple mathematical models to understand the phenomenology, and try afterwards, if possible, to see if the initial model is not too simple and to derive from this new ideas. The simple relaxational model [17] I had imagined allows us at least to understand early observations by Reynolds himself, that do not seem to have attracted much attention, despite their rather spectacular character. The transition to a turbulent state in this kind of model takes place when the turbulent state invades the laminar state, as observed in pipe flows. The transition is defined by the value of the parameter such that the speed of expansion becomes positive. I had predicted too that, near this transition, the turbulent fluctuations "renormalize" the exponents to their value for the directed percolation, an assumption that seems to work rather well in many cases. The same theory explains the formation of turbulent spirals in a Taylor–Couette flow in a subcritical regime [18], where the growth of the turbulent domain brings down the Reynolds number to its "equilibrium" value where the growth stops.

Another prediction specific to non variational models, such as the Navier–Stokes equations, is the possibility of stable localized structures [19] (such types of structures exist as well in variational systems, but they are always unstable there, by an argument of Gibbs). This had been pointed out first by Boris Malomed [20].

Of course, fluid mechanics tries first and foremost to understand what determines the structure of a given flow, given the various forces, such as buoyancy, viscous stresses, inertia, etc. An interesting bifurcation in this set of researches has been the consideration of the so-called passive scalar problems: one assumes a frozen (= time-independent) flow and looks at the behavior of a particle that is convected and that diffuses by Brownian motion. The interesting limit is one of a large Péclet number, where the small molecular diffusion is a singular perturbation. A first result there is that the effective molecular diffusion coefficient is the geometric average of the molecular and turbulent diffusion coefficients [21]. I have always thought that this result could be of practical interest for systems where one needs to increase the contrast between the molecular diffusion coefficients of different molecules to separate them physically. A nice application of the same ideas explained the experiments of Blaise Simon [21]: he observed impurities settling at the bottom of upward-moving separatrices in a cellular flow. Another application of the same kind of idea is to the case of reaction-diffusion equations in frozen cellular flows, a healthy exercise in boundary layer methods [22].

As a last example of this line of developments, I mention the application of these ideas to elasticity problems, quite outside of Bénard interests for sure! For a rather long time it has been known that a flat plate under in-plane pressure buckles out of its plane. If it is long and narrow, the buckling pattern is periodic, very much like the flow field in a system of Rayleigh–Bénard rolls. A first work on this topic [7] had put in evidence the selection of the bifurcated structure by side effects. This had raised my interest for elasticity problems, especially for what I call for lack of better naming, the von Kármán problem. This is essentially the question of finding the buckling pattern far above threshold. This question makes sense in elasticity theory, not in fluid mechanics: there turbulence sets in far above threshold and one is brought back to the (unsolved) problem of turbulence at very large Rayleigh number.

By contrast the elasticity problem, at least in its Hookean limit, follows from a variational principle and makes what one calls a well-posed problem. Almost twenty years had been necessary, as well as the crucial help of Sergio Rica, to solve the von Kármán problem [23]. For large stresses the structure tends to a pattern that is made at large scale of a set of accordion folds. To accomodate the lateral boundary conditions, these folds break into smaller folds near the lateral edges of the plate, and in smaller and smaller folds until they reach a scale where the bending effects take over (in a sense these bending effects introduce the thickness of the plate as a relevant parameter). This was partly inspired by a classical work of Landau on the real ferromagnets where the magnetic domains merge with the surface by forming a cascade of domains to minimize the magnetostatic energy. An interesting new idea in this field is that the minimization should bring in a Cantor-like structure, because there is a choice made locally at every step of the cascade. Moreover, it brings some light too to the effect of the real boundaries, that could have some relevance for turbulent flows, a domain where, as far as I can see, current understanding is still poor.

To conclude, many, if not most of the ideas I just reviewed were born of many discussions with many colleagues. In the first place I put Pierre Bergé whose loss has been so painful to our community of scientists.

References

1. J.C. Loulergue, P. Manneville, and Y. Pomeau, Interface deflection induced by Marangoni effect: An application to infrared/visible image conversion, *J. Phys. (Paris)* **43**, 1967 (1981).
2. C. Normand, Y. Pomeau, and M.G. Velarde, Convective instability: A physicist's approach, *Rev. of Modern Phys.* **49**, 581(1977).
3. P. Bergé, Y. Pomeau, and C. Vidal *L'ordre dans le Chaos: Vers une approche déterministe de la turbulence*, Hermann, Paris (1984).
4. P. Bergé, M. Dubois, P. Manneville, and Y. Pomeau, Intermittency in Rayleigh-Bnard convection, *J. Phys. (Paris) Lett.* **41**, L-341(1980).
5. L.D. Landau and E.M. Lifshitz, *Statistical Physics*, Addison-Wesley, London (1958).
6. J.E. Wesfreid, Y.Pomeau, M. Dubois, C. Normand, and P. Bergé, Critical effects in Rayleigh–Bnard convection, *J. Phys. (Paris)* **39**, 725 (1979).
7. Y. Pomeau, Non-linear pattern selection in a problem of elasticity, *J. Phys. (Paris) Lett.* **42**, L-1,(1981).
8. Y. Pomeau and S. Zaleski, Wavelength selection in 1D cellular structures, *J. Phys. (Paris)* **42**, 515 (1979).
9. Y. Pomeau and P. Manneville, Wavelength selection in axisymmetric cellular structures, *J. Phys. (Paris)* **42**, 1067(1981); Y. Pomeau, S. Zaleski, and P. Manneville, Axisymmetric cellular structures revisited, *Z. Ang. Math.* **36**, 367 (1985).
10. Y. Pomeau, Caustics of non linear waves and related questions, *Europhys. Lett.* **11**, 713 (1990).
11. Y. Pomeau and S. Zaleski, Pattern selection in a slowly varying environment, *J. Phys. (Paris) Lett.* **44**, L-135(1983).
12. Y. Pomeau and P. Manneville, Stability and fluctuations of cellular structures, *J. Phys. (Paris) Lett.* **40**, L-609(1979).
13. P. Manneville, Y. Pomeau, and S. Zaleski, Dislocation motion in cellular structures, *Phys. Rev.* **A27**, 2710 (1983).
14. Y. Pomeau, Effets non adiabatiques dans les structures cellulaires, *Colloque Structures Cellulaires dans les Instabilités*, Gif-sur-Yvette, September 19–23, 1983.
15. P. Manneville and Y. Pomeau, Grain boundary in cellular structures near the onset of transition, *Phil. Mag.* **A48**, 607 (1983).
16. S. Zaleski, Y. Pomeau, and A. Pumir, Optimal merging of rolls near a plane boundary, *Phys. Rev.* **A29**, 366 (1984).
17. Y. Pomeau, Front motion, metastability and subcritical bifurcations in hydrodynamics, *Physica* **23D**, 3 (1986).
18. J. Hegseth, D. Andereck D., F. Hayot, and Y. Pomeau, Spiral turbulence and phase dynamics, *Phys. Rev. Lett.* **62**, 257 (1989).
19. V. Hakim, P. Jacobsen P., and Y. Pomeau, Fronts vs solitary waves in non equilibrium systems, *Europhys. Lett.* **11**, 19 (1990).
20. Appendix A in "Nonlinear Schrodinger and KdV equations with dissipative perturbations", B.A. Malomed, *Physica* **29 D** 155 (1987).
21. Y. Pomeau, Dispersion dans un écoulement en présence de zones de recirculation, *C.R.Ac. Sci.* t.**301**, Série II, 1323 (1985); E. Guyon, Y. Pomeau, J.-P. Hulin, and C. Baudet, Dispersion in the presence of recirculation zones, *Nucl. Phys.* **B** (Proc. Suppl.) 271 (1987); A. Pumir, Y. Pomeau, and W.R. Young, Transitoires dans l'advection-diffusion d'impuretés, *C.R.Ac. Sci.* t.**306**, Série II,741 (1988); W.R. Young, Y. Pomeau, and A. Pumir, Anomalous diffusion of tracer in convection

rolls, *Phys. Fluids* **A 1**, 462 (1989); B. Simon and Y. Pomeau, Free and guided convection in evaporating layers of aqueous solutions of sucrose. Transport and sedimentation of solid particles, *Phys. Fluids* **A 3**, 380 (1991).

22. B. Audoly, H. Berestycki H., and Y. Pomeau, Réaction diffusion en écoulement stationnaire rapide, *C.R. Ac. Sci.*, t.**328**, Série IIb, 255 (2000).

23. Y. Pomeau and S. Rica, Plaques très comprimées, *C.R. Ac. Sci.*, t.**325**, Série II,181 (1997).

6 Bénard Convection and Geophysical Applications

Friedrich H. Busse

Institute of Physics, University of Bayreuth, D-95440 Bayreuth, Germany

Some typical relationships between experimental and theoretical studies of Rayleigh–Bénard convection and convection phenomena observed in geophysical and astrophysical systems are discussed. Because of the vast range of the subject only a few examples are described in a qualitative manner. Convection in planetary cores and its dynamo action receives special attention.

6.1 Introduction

The regular cellular fluid flows of Bénard convection have long been considered a peculiar phenomenon without more general relevance. As can still be read in older books (see, for instance, [1]) Bénard cells were regarded as an interesting curiosity, but without relationship to fundamental questions such as the transition to turbulence in fluid systems. The transition to complex flows such as turbulence was viewed as a stochastic process and it was not even generally accepted that the Navier–Stokes equations are capable of describing this process. Beginning in the 1960s in the last century this view changed entirely. The transition from the static basic state to Bénard cells is now regarded as the primary example for the transition from simple to more complex states of fluid flow, and the direct transitions to turbulent states observed in pipe or channel flows must now be considered as special cases due to their complex subcritical bifurcation structures. Because in the Rayleigh–Bénard problem the onset of convection occurs in the form of supercritical or only weakly subcritical bifurcations, the flow is laminar and the breaking of the symmetry of the basic state is not associated with the onset of time dependence and random behavior. This property can also be observed in higher transitions, and steady cellular type flows can be realized in experiments up to 500 times the critical value of the Rayleigh number [2]. It is thus not surprising that Rayleigh–Bénard convection has become the primary example for the study of sequences of bifurcations in nonlinear fluid dynamics. Here and in the following we refer to Rayleigh–Bénard convection whenever thermal buoyancy is the primary source of mechanical energy. As is well known [3], Bénard originally did not realize that the temperature dependence of surface tension was the origin of the cellular motion that he observed. Surface-tension-driven flows are now referred to as Bénard–Marangoni convection.

The large family of Rayleigh–Bénard systems and its extensions have been indicated in Figure 6.1. Only systems that are essentially isotropic in the hor-

Convection in the Presence of (nearly) Two-Dimensional Isotropy Under Steady External Conditions

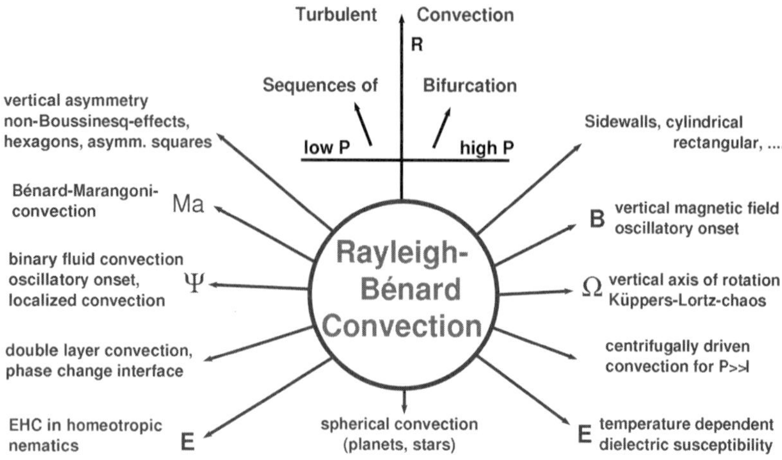

Fig. 6.1. Rayleigh–Bénard convection and its extensions to fluid systems that are nearly isotropic in two dimensions.

izontal plane have been included. Many more systems exhibit similar scenarios of supercritical onset of convection and sequences of bifurcations, but include a preferred direction such as that of a shear flow superimposed on the convection layer or in the case of an inclined convection layer. The Taylor–Couette system (see, e.g., the review of DiPrima and Swinney [4], or the contribution of Prigent et al. Chapter 13 in this volume) should also be mentioned in this connection, inasmuch as, it is another well-known system that exhibits sequences of bifurcations and a delayed onset of turbulence.

Among the examples listed in Figure 6.1 quite a few are related to geophysical phenomena and geophysical and astrophysical observations have often provided the motivation for research on Rayleigh–Bénard convection. Bénard and his students were well aware that cumulus clouds represent a convection phenomenon and in laboratory experiments they demonstrated that cloud streets arise from the alignment of convection rolls with shear of a mean wind [5, 6]. These experiments were among the first of a long sequence of laboratory studies of dynamical phenomena occurring in the atmosphere which continue to this day. The fact that a close relationship exists between cellular structures realized in a laboratory experiment and those observed in a highly turbulent system such as the atmosphere is rather surprising, especially for scientists brought up on the view that turbulence is just a random process that can be studied only with statistical tools. Convection cells in the Earth's atmosphere and on the sun are particularly striking examples for coherent structures in turbulent systems and support the notion that in a first approximation the effects of turbulence can be taken into account in the form of an appropriate eddy viscosity.

In the following we first briefly survey geophysical and astrophysical convection phenomena in Section 6.2 and then return in Section 6.3 to some exemplary cases to be considered in detail. A particular subject of current intense research is the generation of magnetic fields by convection in planetary cores and in stars which is discussed in Section 6.4. The chapter concludes with an outlook on future research.

6.2 Convection in Geophysical and Astrophysical Systems

6.2.1 Thermal Convection in the Atmosphere

Convection cells occur when a horizontal fluid layer of thickness d is heated from below and the density difference between upper and lower parts becomes sufficiently strong such that the potential energy gained by overturning motion can overcome the losses due to thermal and viscous dissipation.

Fig. 6.2. Cloud bands caused by convection rolls that form when cold oceanic air moves over heated land.

This requirement is expressed mathematically in the form

$$R \geq R_c \quad \text{with} \quad R \equiv \frac{\alpha(T_2 - T_1)gd^3}{\nu\kappa}, \tag{6.1}$$

where α is the coefficient of thermal expansion, T_2 and T_1 are the temperatures at the lower and upper boundaries of the fluid layer, g is the acceleration of gravity, and ν and κ are the kinematic viscosity and the thermal diffusivity of the fluid, respectively. Besides the dimensionless parameter R, called the Rayleigh number, an additional parameter, the Prandtl number P, is necessary to characterize the properties of the fluid layer. P denotes the ratio ν/κ and enters the description of convection usually only in the case of finite amplitude convection beyond the critical value R_c of R. R_c assumes the value 1708 when rigid boundaries with fixed temperatures are used.

Observations of convection cells should be a common experience of daily life inasmuch as fluids are heated from below in pots or pans and in puddles of water in the streets that are cooled from above by evaporation. But these motions are not easily discerned and some method of visualization must be employed. Bénard [7] used fine graphite powder, which he sprinkled on the surface of a layer of molten spermaceti which he heated from below. Convection in the atmosphere is easily visible when clouds are present. As the moist air in the rising part of motion cools on its way up, water droplets appear owing to condensation and clouds are formed. Here and in other cases where the hydrostatic pressure varies significantly between the top and bottom of the convection layer it is important to identify $T_2 - T_1$ in expression (6.1) with the superadiabatic part of the temperature distribution. Otherwise the atmosphere would always be convectively unstable according to criterion (1) because the temperature drops by 6°C per kilometer height on average. Because a parcel of air is cooled by expansion on its way up, it can keep its buoyancy only if the surrounding air is even cooler, that is, if the temperature gradient exceeds the adiabatic one in absolute values. Figure 6.2 shows typical convection rolls in the atmosphere made visible by clouds. Even the dislocation generated when two pairs of rolls are joined into a single pair can be seen just in the same way as in a laboratory experiment (see, e.g., [8, 9]).

Since the advent of satellite observations, the highly regular structure of large-scale convection phenomena has become apparent. Cells of a typical diameter of 30 km are made visible by the cloud patterns as seen, for example, in Figure 6.3. These structures are known as mesoscale convection in the meteorological literature and have been the subject of intense research during the past decades. For a recent review we refer to Atkinson and Zhang [10]. Although there is little doubt that mesoscale convection is caused by thermal buoyancy, the large horizontal extent of the cells that exceeds the height of the participating atmospheric layer by a factor of 20 or more is still not well understood. Another interesting feature of mesoscale convection is the property that "open" and "closed" cells can be distinguished. When the air in the middle of the cells is descending, the water droplets tend to evaporate and the sky becomes clear or "open". The reverse situation occurs in closed cells. A similar distinction has long

Fig. 6.3. Mesoscale convection visualized by cloud patterns with open and closed cells photographed by the crew of the Shuttle Endeavour.

been observed in laboratory experiments where rising motions in the center of the hexagonal cells are usually observed when convection occurs in liquids and the opposite direction is found in convecting gases. These findings have been attributed to the opposite temperature dependence of the viscosity in liquids and in gases [11]. Experiments of V. Tippelskirch [12] on convection in molten sulphur, which reverses the temperature dependence of its viscosity, confirmed this interpretation. It should be kept in mind that material properties other than viscosity have a similar influence through their dependence on the temperature [13], but the temperature dependence of viscosity is usually the strongest. Through the choice of the flow direction such that the viscosity is lowest where the motion of highest strain occurs, namely, in the center of the hexagonal cell, a state of minimal dissipation at given amplitude of convection is achieved. This interpretation is confirmed by the stability analysis for the competing hexagonal patterns [13].

Different temperature dependences of material properties are not likely to explain the preference for open or closed cells in mesoscale convection in the atmosphere, inasmuch as both types of cells are often observed together. In this connection the observations by Assenheimer and Steinberg [14] of coexisting upflow and downflow hexagons in a fluid with nearly constant material properties are of interest. Indeed, a theoretical analysis of the stability of hexagonal convection patterns has confirmed that both types of hexagonal convection become stable in the experimentally relevant regime of Rayleigh–Bénard numbers of the order of 10^4 [15]. This result indicates that the direction of motion in hexagonal convection cells is not always a good indicator of properties of the fluid and may sometimes just depend on initial conditions.

6.2.2 Some Remarks on Mantle Convection

Thermal convection is a ubiquitous source of motion in all fluid geophysical systems besides the atmosphere such as lakes and oceans or the liquid outer core of the Earth. But even in the mantle of the Earth which is usually regarded as solid because it transmits shear waves, slowly moving convection flows are of fundamental importance because they break up continents and form mountain ranges. The plate tectonic "revolution" in the geosciences around 1960 has validated the early hypothesis that the continents are moving [16] and that the mantle is convecting [17] and has led to extensive investigations of large-scale flows in the Earth beneath our feet. If the mantle would convect like a laboratory fluid layer heated from below, steady convection cells may be expected because the Prandtl number is huge, $P \approx 10^{22}$. The analogue of hexagonal cells would look like the patterns shown in Figure 6.4, depending on the radius ratio of the participating mantle layer. Because the lighter continents cannot be subducted into the mantle in contrast to the heavier, basaltic oceanic crust, the boundary conditions for mantle convection differ strongly from those in the usual laboratory experiments. Phase transitions of the mantle material at depths of 400 km and of 660 km also influence the dynamics of the mantle such that a strongly time-dependent process results. It has to be kept in mind, of course, that the typical time scale of mantle convection is counted in hundreds of millions of years and that the continuing cooling of the Earth prevents the attainment of a steady state anyhow.

There is considerable evidence from seismic tomography [20, 21] that a mode of the quadrupolar form $l = 2$ such as that shown in the upper left corner of Figure 6.4 dominates in mantle convection. The axis of this mode is not aligned with the axis of rotation of the Earth as one might expect because of the equatorial bulge of the Earth. Instead the axis lies in the equatorial plane. It turns out that the equatorial bulge of the Earth has little effect on mantle convection.

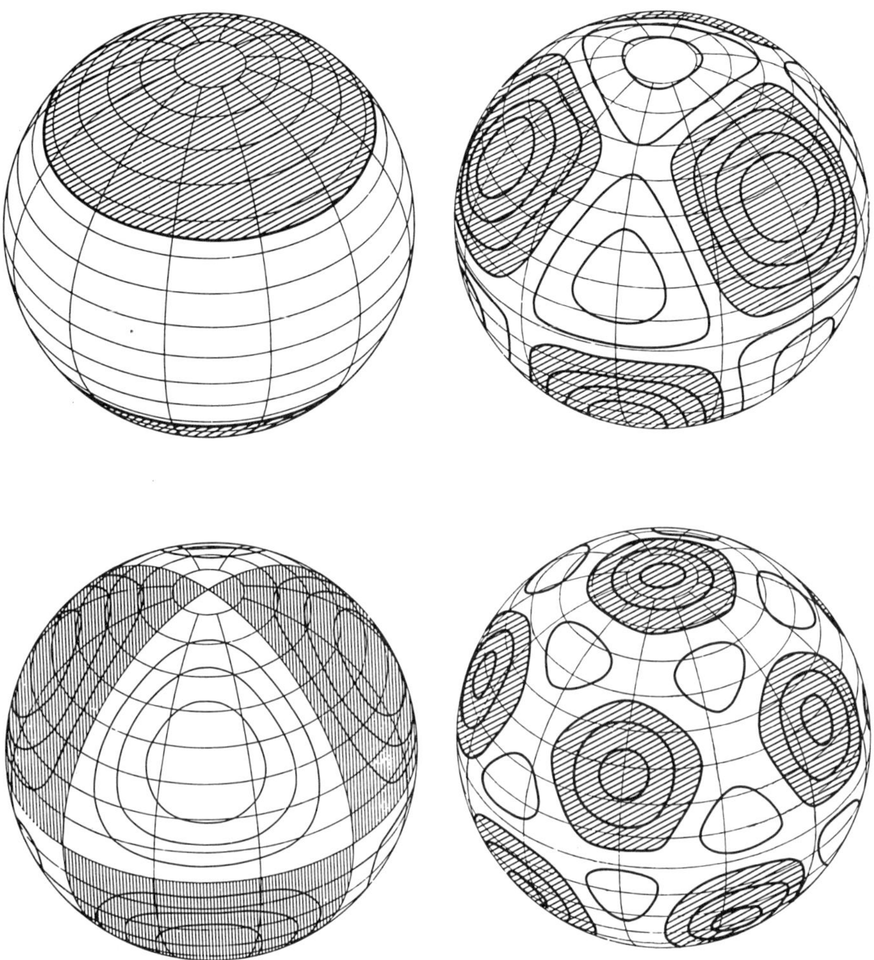

Fig. 6.4. Preferred convection patterns in spherically symmetric fluid shells. The patterns correspond to the degrees $l = 2, 4$ (upper row) and $l = 3, 6$ of the spherical harmonics [18, 19].

It adjusts itself relatively quickly to the instantaneous axis of rotation. On the contrary, it is the quadrupolar ($l = 2$) component of mantle convection that determines the position of the axis of rotation. Because the latter coincides in the time average with the axis of maximum moment of inertia and this in turn is most strongly influenced by the dynamical effects of the quadrupolar component of convection, the axis of rotation orients itself perpendicular to the axis of the quadrupolar mode [22]. This feature is reflected in the corresponding deviations of the geoid and of the gravitational potential of the Earth from their axisymmetric forms and also in the property of the predominantly north–south extension of the Atlantic and Pacific ocean basins.

These latter properties have long puzzled geophysicists because the Coriolis force cannot have any effect on mantle convection because of the high viscosity μ of the order of $10^{22} Pa\,s$. The corresponding dimensionless parameter τ, also called the Coriolis number, is defined by $\tau = 2\Omega d^2 \varrho/\mu$, where Ω is the angular velocity of the Earth, d is the thickness of the mantle, and ϱ is its average density which is roughly $5 \cdot 10^3 \mathrm{kg}\ \mathrm{m}^{-3}$. Based on these values $\tau \approx 10^{-10}$ is found which clearly indicates that only through the effect of the $l = 2$ convection component on the moment of inertia of the Earth can a connection between mantle convection and the axis of rotation be established.

There are many more aspects of mantle convection that can be related to Bénard cells, such as the influence of varying viscosity or the influence of internal heat sources of the fluid. We refer to the monographs by Peltier [23] and Schubert et al. [24] where the subject of mantle convection is treated in detail.

6.2.3 Solar Convection

The granular appearance of the solar surface was already noticed by the British astronomer William Herschel in 1801, but it was not associated with thermal convection at that time. A hundred years later much more detailed observations of the solar photosphere had become available and a connection between solar granulation and thermal convection was discussed. Soon after he had published his first observations, Henri Bénard was made aware of the similarity between the convection patterns he had seen and the solar granulation photographed by Janssen [25, 6]. The interpretation of the solar granulation as a convection phenomenon was confirmed by theoretical considerations in the 1930s when it became evident that the heat transport in the outer part of the sun could not be carried solely by radiation and that the photosphere is indeed unstable with respect to convective instabilities.

Today four separate scales of solar convection are distinguished. Convection cells corresponding to the solar granulation have typical diameters of up to 2000 km and are highly time dependent such that their typical lifetimes are only about 5 minutes. A photograph of solar granulation is shown in Figure 6.5 which also shows a sunspot. Convection is suppressed by the strong vertical magnetic field in the dark umbra of the sunspot. In the surrounding penumbra, convection rolls aligned with the horizontal component of the magnetic field are visible. Mesoscale convection corresponding to cells with a diameter of 5000 to 10^4 km have been identified only rather recently through observations of Doppler shifts of photospheric spectral lines. The lifetimes of mesoscale cells may be as long as a few hours. Supergranular cells can also be noticed through Doppler shifts of spectral lines. But in contrast to the vertical component of the velocity observed in the case of mesoscale cells, supergranulation is best seen through the Doppler shifts caused by the horizontal component of the velocity field on the sun. The size of supergranular cells is about $3 \cdot 10^4$ km and their lifetimes are of the order of one day. All of these three types of convection have in common that the cells are

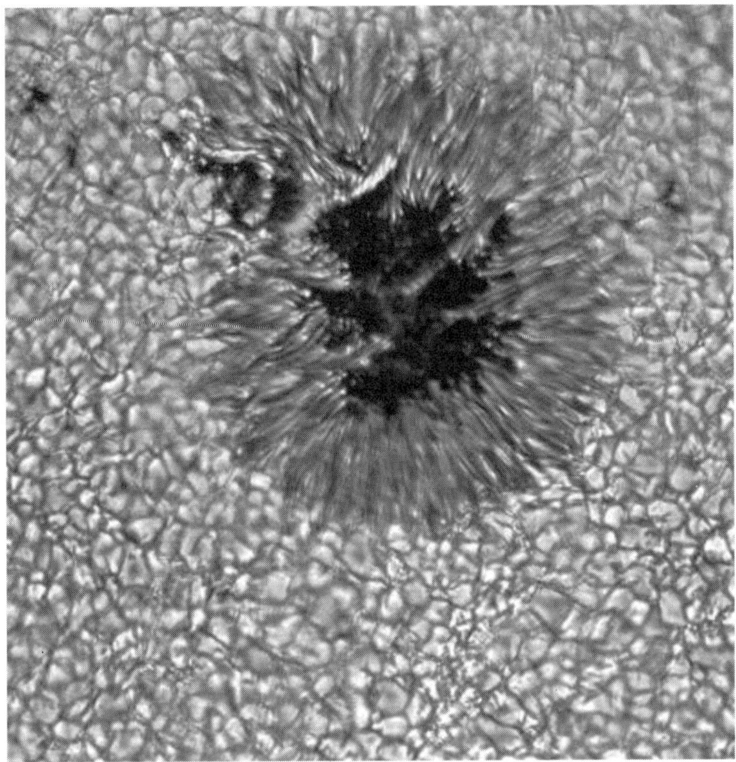

Fig. 6.5. Photograph of solar granulation near a sunspot (from T. Rimmele/NSO/AURA/NSF).

polygonal and that the gas rises in the center, moves outward, and descends at the rim of the cells. There is less observational evidence available for convection in the form of giant cells that are supposed to extend throughout the entire solar convection zone and thus correspond to a size between 10^5 and $2 \cdot 10^5$ km. Because of their large size and their expected lifetime on the order of the period of solar rotation, giant cells will assume the form of "banana" cells elongated in the direction of the axis of rotation [26]. We return to the effects of rotation in Section 6.3.

Convection at the surface of the sun represents the most turbulent system accessible to direct observations. It is thus rather surprising that fairly discrete scales should characterize this system rather than a broad spectrum of wavenumbers. A first theoretical attempt at an explanation of the discrete scale was made by Simon and Weiss [27], but mesoscale convection was not known at that time. The depths at which hydrogen and helium become ionized play a role. But it should be kept in mind that turbulent convection transports heat most effectively if it is done on discrete scales. Perhaps the solar convection zone exhibits the discrete scales differing by about a factor of four which are predicted for an idealized vector field optimizing the heat transport at an asymptotically high Rayleigh number [28].

6.3 Convection in Rapidly Rotating Fluid Spheres

The problem of convection in rotating spherical fluid shells is a fundamental dynamical problem in geophysics and astrophysics. Many properties of celestial bodies such as the magnetic fields of the Earth and other planets and of stars or the band structures observed on the major planets depend on the dynamics of convection in the interior of these systems. Moreover, it is important to know about the heat transport in these systems in order to understand their evolution in time. Not surprisingly, an extensive literature has been devoted to convection in rotating spherical systems and we touch here only upon some basic features.

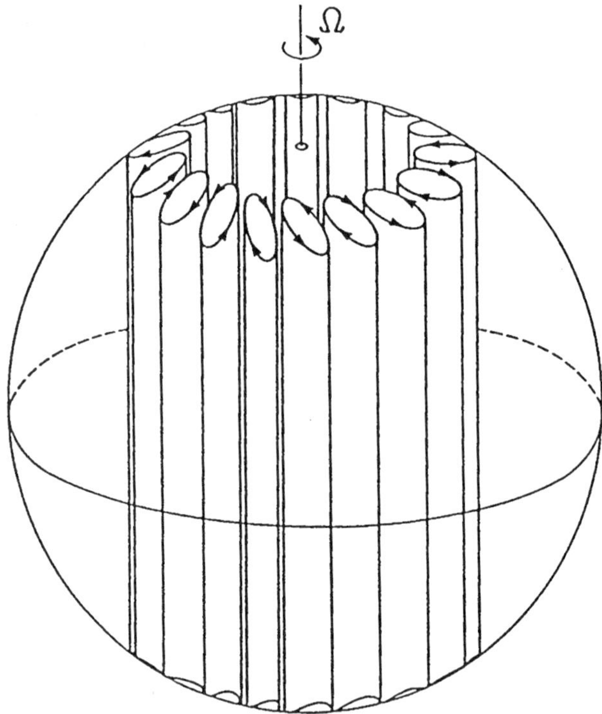

Fig. 6.6. Sketch of convection columns in an internally heated rapidly rotating, self-gravitating fluid sphere.

Because the Coriolis force is the dominating force in the equations of motion relative to the rotating frame of reference it can be balanced only by the pressure gradient unless one is considering motions in the form of inertial waves. But the latter are rather inefficient in transporting heat and become preferred at onset of convection only at very low Prandtl numbers. The balance between Coriolis force and pressure gradient,

$$2\boldsymbol{\Omega} \times \varrho\boldsymbol{v} \approx -\nabla p, \tag{6.2}$$

where $\boldsymbol{\Omega}$ is the angular velocity vector of rotation and ϱ is the density, also applies to the atmosphere and has been christened "geostrophic balance" by meteorologists. A consequence of relationship (6.2) is the Proudman–Taylor theorem,

$$\boldsymbol{\Omega} \cdot \nabla \varrho\boldsymbol{v} \approx 0, \tag{6.3}$$

which is obtained after the curl of (6.2) is taken and the equation of continuity in the form $\nabla \cdot \varrho\boldsymbol{v} = 0$ is used. Equation (6.3) states that the momentum vector $\varrho\boldsymbol{v}$ must not depend on the coordinate in the direction of the axis of rotation when the geostrophic balance is valid. Of course, in a three-dimensional world velocity fields cannot obey the theorem in general. The boundary conditions in a rotating spherical fluid shell alone prevent the exact validity of property (6.3). But there is a tendency of the realized convection modes to approach this property as is evident from the sketch of the asymptotic form of convection in a rotating, internally heated fluid sphere as shown in Figure 6.6.

To understand the dynamics of convection in rotating spheres in more detail it is convenient to consider the model of the rotating cylindrical annulus. As sketched in Figure 6.7 the annular configuration with conical end surfaces can be regarded as a cylindrical cut from the sphere including the region occupied by the convection flow in Figure 6.6. Starting with a velocity field that is geostrophic in first approximation

$$\boldsymbol{v} = \nabla\psi(x, y, t) \times \boldsymbol{k} + \ldots, \tag{6.4}$$

we obtain the following equation for the z-component of the vorticity, $\boldsymbol{k} \cdot \nabla \times \boldsymbol{v} = -\Delta_2\psi$,

$$\frac{\partial}{\partial t}\Delta_2\psi + \boldsymbol{v} \cdot \nabla\Delta_2\psi + \tau\boldsymbol{k} \cdot \nabla\boldsymbol{v} \cdot \boldsymbol{k} = -R\Delta_2\Theta + \nabla^2\Delta_2\psi, \tag{6.5}$$

where Δ_2 is the two-dimensional Laplacian, $\Delta_2 = \partial^2/\partial x^2 + \partial^2/\partial y^2$. We have used dimensionless variables in Equations (6.4) and (6.5) based on the gap width D, the time scale D^2/ν, and temperature scale $T_2 - T_1$. Here ν denotes the kinematic viscosity of the fluid, and T_1 and T_2 are the temperatures kept fixed at the inner and outer cylindrical boundary of the annulus, respectively. A Cartesian coordinate system x, y, z has been introduced with x in the radial direction, y in the azimuthal direction, and z in the direction of the axis of rotation parallel to the unit vector $\boldsymbol{k}$ because we are assuming the small gap approximation. We also use the Boussinesq approximation in which the temperature dependence $\varrho = \varrho_0[1 - \alpha(T_2 - T_1)(x/D + \Theta)]$ of the density is taken into account only in the gravity term and all material properties are regarded as constants otherwise. In applications to convection in planetary cores, gravity would point radially inward and T_1 would exceed T_2. But in laboratory realizations of the problem

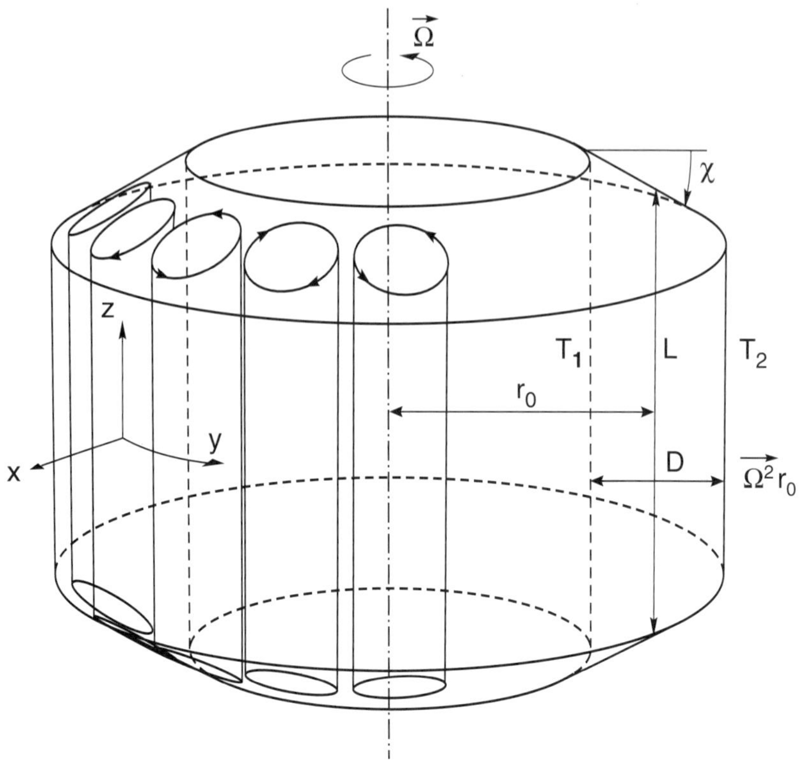

Fig. 6.7. Geometrical configuration of the rotating cylindrical annulus.

centrifugal force is used as effective gravity, $g = \Omega^2 r_0$, where r_0 is the mean radius of the annulus, and the temperature gradient must be reversed, $T_1 < T_2$. Because only the product of gravity and applied temperature gradient enters, the two cases are mathematically identical. We have written Equation (6.5) for the laboratory case such that the Rayleigh number R given by (6.1) is positive.

Because the velocity field (6.4) is z-independent in first approximation, all terms in Equation (6.4) are reproduced when it is averaged over z except for the term proportional to $\tau = 2\Omega D^2/\nu$. Because of the z-derivative this term can be replaced by

$$2\tau \frac{D}{L} tg\chi \frac{\partial \psi}{\partial y} \equiv \eta \frac{\partial \psi}{\partial y} \tag{6.6}$$

after the boundary condition

$$v_x \sin \chi \pm v_z \cos \chi = 0 \quad \text{at } z = \frac{1}{2}\frac{L}{D} \tag{6.7}$$

at the conical top and bottom boundaries has been used. Of course, the angle χ must be small, such that the assumption of a small nongeostrophic component of the velocity field $\boldsymbol{v}$ is justified. Indeed, in the limit $\chi \to 0$ the velocity field is entirely geostrophic if no viscous stresses are exerted by the upper and lower boundaries. In the presence of rigid boundaries, Ekman layers are formed that give rise to a small z-component v_z of $\boldsymbol{v}$ of the order $\tau^{-1/2}$ [29].

We have assumed that the deviation Θ of the temperature field from the basic state of pure conduction is also independent of z. This property is satisfied if the thermal conductivity of the conical boundaries is low compared to that of the fluid. Θ thus obeys the equation

$$P\left(\frac{\partial}{\partial t} + \frac{\partial}{\partial y}\psi\frac{\partial}{\partial x} - \frac{\partial}{\partial x}\psi\frac{\partial}{\partial y}\right)\Theta - \Delta_2\Theta = \frac{\partial}{\partial y}\psi. \tag{6.8}$$

This equation together with Equation (6.5), which can now be written in the form

$$\left(\frac{\partial}{\partial t} + \frac{\partial}{\partial y}\psi\frac{\partial}{\partial x} - \frac{\partial}{\partial x}\psi\frac{\partial}{\partial y} - \Delta_2\right)\Delta_2\psi - \eta\frac{\partial}{\partial y}\psi = R\frac{\partial}{\partial y}\Theta, \tag{6.9}$$

can be easily solved in the case when stress-free conditions can be assumed at the cylindrical walls, $x = \pm\frac{1}{2}$, and when the amplitude of convection is sufficiently small such that nonlinear terms can be neglected. Using the ansatz $\psi, \Theta \sim \sin n\pi(x + \frac{1}{2})\exp\{i(qy + \omega t)\}$ and using Equation (8a) to eliminate Θ from Equation (8b) we obtain an algebraic equation, the real and imaginary parts of which determine the values of ω and R for which a solution with given q exists:

$$\omega = \frac{-\eta q}{(1 + P)[((n\pi)^2 + q^2]} \ , \quad R = \frac{[(n\pi)^2 + q^2]^3}{q^2} + \frac{2\eta_p^2}{(n\pi)^2 + q^2} \tag{6.10}$$

where the definition $\eta_p \equiv \sqrt{2}P\eta(1 + P)^{-1}$ has been used.

The result (6.10) indicates that convection columns propagate as do Rossby waves in the prograde (retrograde) direction when the height of the gap decreases (increases) with increasing distance from the axis of rotation. Because of this property this kind of convection has been called " thermal Rossby waves." In the limit $\eta = 0$ the relationship for Rayleigh–Bénard convection is recovered in which case it is well known that the mode with $n = 1$ corresponds to the lowest value of R and thus describes the onset of convection. This property also holds for finite values of η. But as the limit of large η is approached, the minimizing Rayleigh number becomes independent of n in first approximation,

$$q_c = \eta_p^{1/3}(1 - \frac{7}{12}(n\pi)^2\eta_p^{-2/3} + \ldots), \omega_c = -\sqrt{2}P^{-1}\eta_p^{2/3}(1 - \frac{5}{12}(n\pi)^2\eta_p^{-3/2} + \ldots)$$

$$R_c = \eta_p^{4/3}(3 + (n\pi)^2\eta_p^{-2/3} + \ldots). \tag{6.11}$$

The property that the radial structure is of secondary importance has important consequences:

(1) The nature of the cylindrical boundaries has little influence on the thermal Rossby waves. This is the main reason that the annulus model can be applied to the spherical case where the cylindrical boundaries are absent.

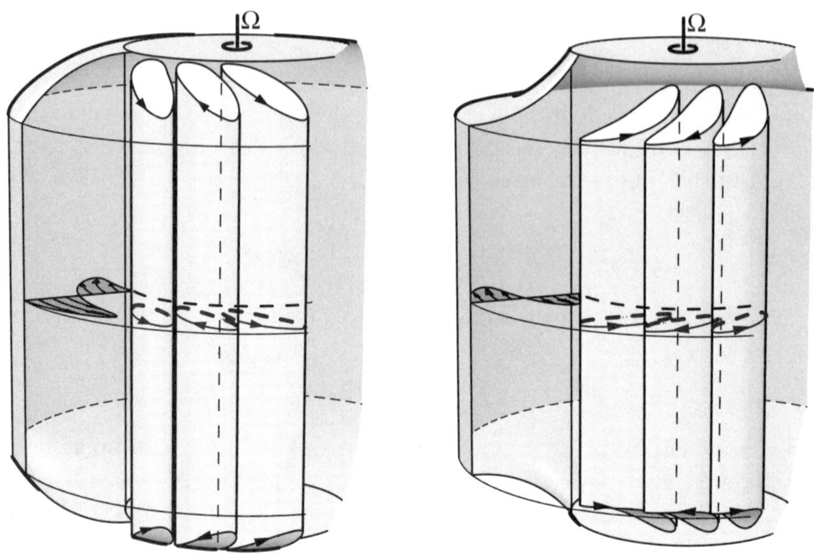

Fig. 6.8. Influence of curved conical boundaries on convection columns. In the convex case (a), the columns tend to spiral outward in the prograde direction creating thereby a differential rotation as indicated. The opposite situation is found in the concave case (b).

(2) Modes corresponding to different values of n compete at supercritical Rayleigh numbers which leads to bifurcations from the state of simple thermal Rossby waves.

An exact solution of equations (6.8, 6.9) is no longer possible when the conical boundaries are curved as shown in Figures 6.8a,b. The x- and y-dependences are no longer separable if the parameter η is replaced by a function of x, say $\eta = \eta_0(1 + \epsilon x)$. In the case of a convex boundary as in Figure 6.8a, we must expect that the convection phase increases radially outward because the thermal Rossby waves tend to propagate faster in the outer part than in the inner part of the gap. The opposite situation occurs in the case of concave boundary ($\epsilon < 0$) as shown in Figure 6.8b. The most important consequence of the resulting tilt of the convection columns is that an angular momentum transport is induced owing to the Reynolds stress $\overline{v_x v_y}$ where the bar indicates the y-average. As a consequence a differential rotation is generated as indicated in Figure 6.8a and b. The two kinds of differential rotation with opposite signs have indeed been observed in laboratory experiments [30].

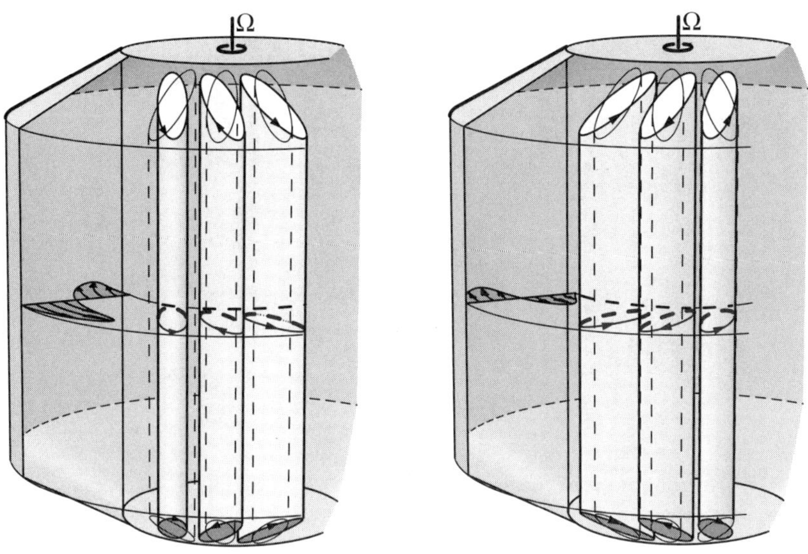

Fig. 6.9. The mean flow instability leading to either outward (a) or inward (b) transport of prograde angular momentum.

There is another process that also causes differential rotation and which appears to be even more important in geophysical and astrophysical applications of the theory. The mean flow instability of thermal Rossby waves causes a spontaneous tilt of the thermal Rossby waves as indicated in Figures 6.9a and 6.9b. Because the differential rotation generated by the tilt tends to reinforce the latter, an exponential growth of the instability occurs until a new equilibrium is reached with asymmetric convection columns that are either stronger in the inner or in the outer part of the gap. Both signs of the instability are equally likely in the absence of curvature of the conical boundaries. But because the spherical case is characterized by a convex curvature we expect that in this case an equatorial acceleration will occur as is indeed observed.

The mean flow instability [31, 32] basically results from property (6.2) mentioned above. A perturbation with a $\sin 2\pi(x + \frac{1}{2})$-dependence grows and yields a Reynolds stress of positive or negative sign depending on the phase of the perturbation relative to the stationary Rossby wave. With increasing Rayleigh number, the differential rotation tends to grow even faster than the amplitude A of convection because the Reynolds stress is proportional to A^2. In the case of spherical convection this leads to the phenomenon of " localized convection" where the convection columns are sheared off in the major part of the spherical

Fig. 6.10. Regular (a) and localized turbulent convection (b) in a rotating spherical fluid shell with $\tau = 10^4$. Dark and light surfaces correspond to a fixed value of the radial component of the velocity with positive and negative sign, respectively. Rayleigh and Prandtl numbers are $R = 3.8 \cdot 10^5, P = 10$ (a) and $R = 7 \cdot 10^5, P = 1$ (b).

shell as indicated in Figure 6.10b. Only in a smaller part convection still occurs and continues to drive the differential rotation. For further details see Grote and Busse [33].

At even higher Rayleigh number values this delicate balance between the differential rotation and convection is no longer possible and relaxation oscillations set in. Once a strong differential rotation has been generated it shears off the convection columns and convection dies. But without convection the differential rotation can no longer be sustained and must decay owing to viscous dissipation. As soon as the differential rotation has decayed sufficiently, convection columns start to grow again and a sharp growth of the differential rotation occurs almost simultaneously. As a result, the cycle thus repeats itself as illustrated in Figures 6.11 and b. It is remarkable how regular the relaxation oscillations occur inasmuch as at this high Rayleigh number convection has long become turbulent.

6.4 Convection-Driven Dynamos in Spherical Fluid Shells

The generation of planetary and stellar magnetic fields by convection flows in the electrically conducting fluid interiors of these celestial bodies is one of the most fascinating subjects of geophysics and astrophysics. Since Larmor [34] proposed the dynamo process as an explanation for the intense magnetic fields

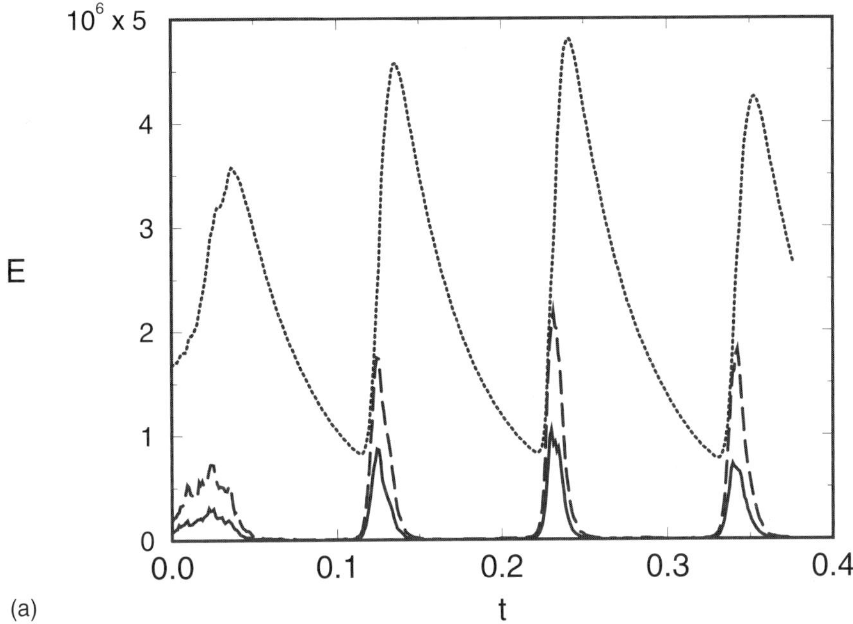

Fig. 6.11. (a) Relaxation oscillations of turbulent convection in the case $\tau = 1.5 \cdot 10^4$, $R = 1.2 \cdot 10^6$, $P = 0.5$. The dotted, dashed, and solid lines indicate the energies of the differential rotation and of the toroidal and poloidal components of the nonaxisymmetric convection columns as a function of time (for details, see [33]).

observed in sunspots, the induction of magnetic fields by motion in an electrically conducting fluid has become an flourishing field of research. During the first half of the twentieth century the dynamo process of the generation of magnetic fields in simply connected regions of essentially uniform electrical conductivity still appeared doubtful, because it seemed unlikely to many physicists that an electromotive force could be generated by fluid flows without a short-circuiting within the electrically conducting domain. Indeed, Cowling [35] proved in 1934 that axisymmetric magnetic fields could not be generated by the dynamo process. Through the dynamo models created by Backus [36] and Herzenberg [37], however, it was demonstrated in a mathematically convincing manner that the dynamo process is feasible.

Although in earlier work on dynamo theory attention was focused on the kinematic dynamo problem which is concerned with the possibility of a growth of a magnetic field in the presence of an arbitrarily chosen solenoidal vector field as velocity field, it became evident that planetary and stellar dynamos can be understood only if physically feasible velocity fields are used. The problem of convection in planetary cores and in stars must thus be solved first, because flows driven by thermal or chemical buoyancy are the most likely energy source for the magnetic field. The onset of dynamo action can then be considered as an instability which in contrast to the usual hydrodynamic instabilities exhibits the

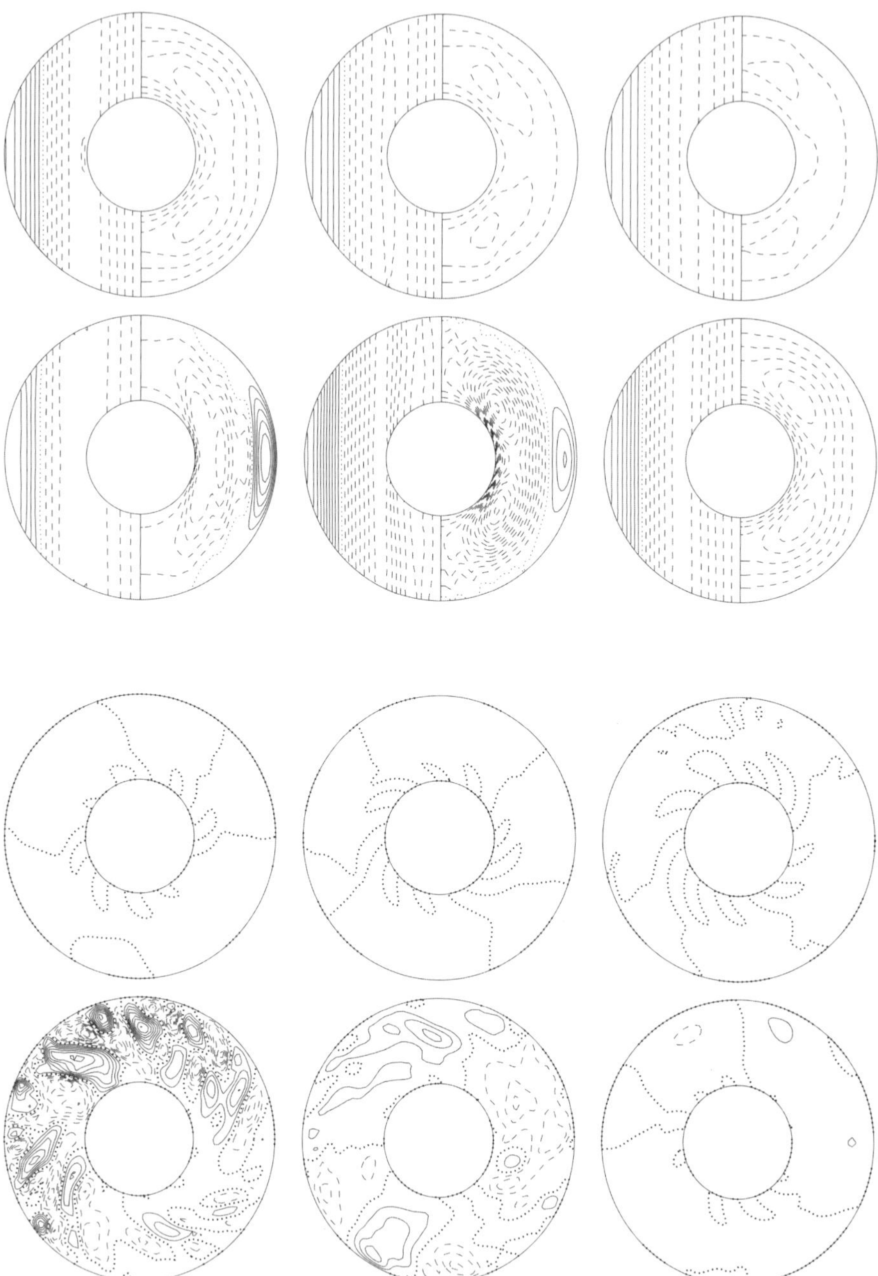

Fig. 6.11. (b) Time series of plots (left to right in first row, then second row with $\Delta t = 0.02$) for the case of (a). The upper part shows lines of constant mean azimuthal velocity $\bar{u}_\varphi$ in the left half and streamlines of meridional circulation in the right half of the circles. The lower part shows streamlines of convection columns in the equatorial plane.

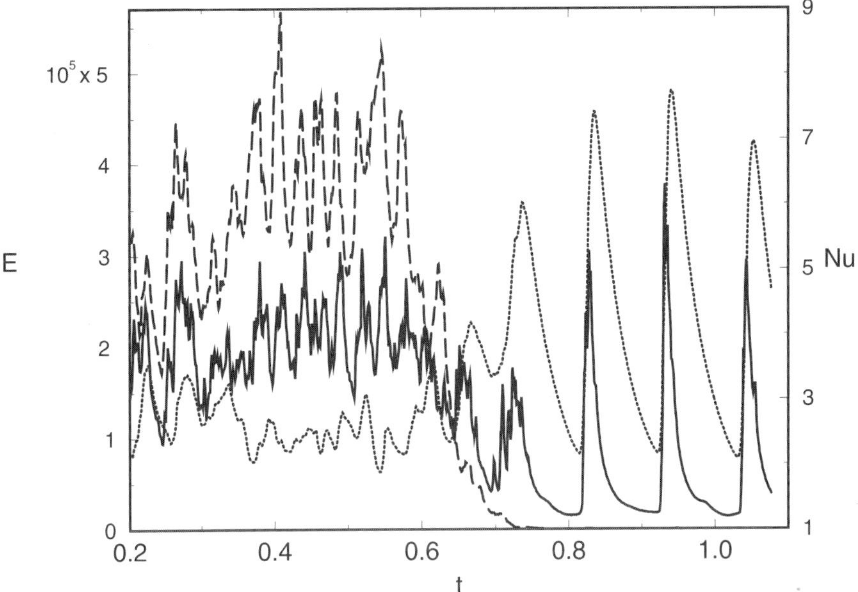

Fig. 6.12. Transition from a dynamo convection state to a relaxation oscillation state in the case $\tau = 1.5 \cdot 10^4$, $R = 1.2 \cdot 10^6$, $P = 0.5$, $P_m = 0.5$. Magnetic energy (multiplied by 7.5, dashed line), energy of differential rotation (dotted line), and Nusselt number (solid line) are plotted in dependence on time t.

property that a new physical quantity, namely, the magnetic field, is associated with it. The usual methods developed for the analysis of bifurcating solutions in hydrodynamic problems can thus also be applied in dynamo theory; see, for example, Childress and Soward [38], Soward [39], or Busse [40, 41]. For more realistic simulations of convection-driven dynamos in rotating spherical fluid shells numerical methods must be used such as the pseudo-spectral method described by Glatzmaier [42]. For recent reviews of this rapidly expanding field of planetary and stellar dynamos and for the geodynamo in particular we refer to the articles by Busse [43], Dormy et al. [44], Fearn [45], and Roberts and Glatzmaier [46].

A typical demonstration of the interaction of convection in a rotating spherical shell and the magnetic field is shown in Figure 6.12. Because the dynamo process could be sustained only marginally by convection in this particular case, the chaotic dynamo decays after a while and convection returns to the state of relaxation oscillations that was described in the preceding section. We thus see the energy balances of convection in both states, with dynamo action and without any significant magnetic field. In the former state the differential rotation is suppressed through the braking effect of the Lorentz force. The amplitude of convection is thus relatively high and the heat transport as measured by the Nusselt number exceeds that realized in the absence of a magnetic field by an order of magnitude. This type of interaction must be distinguished from the re-

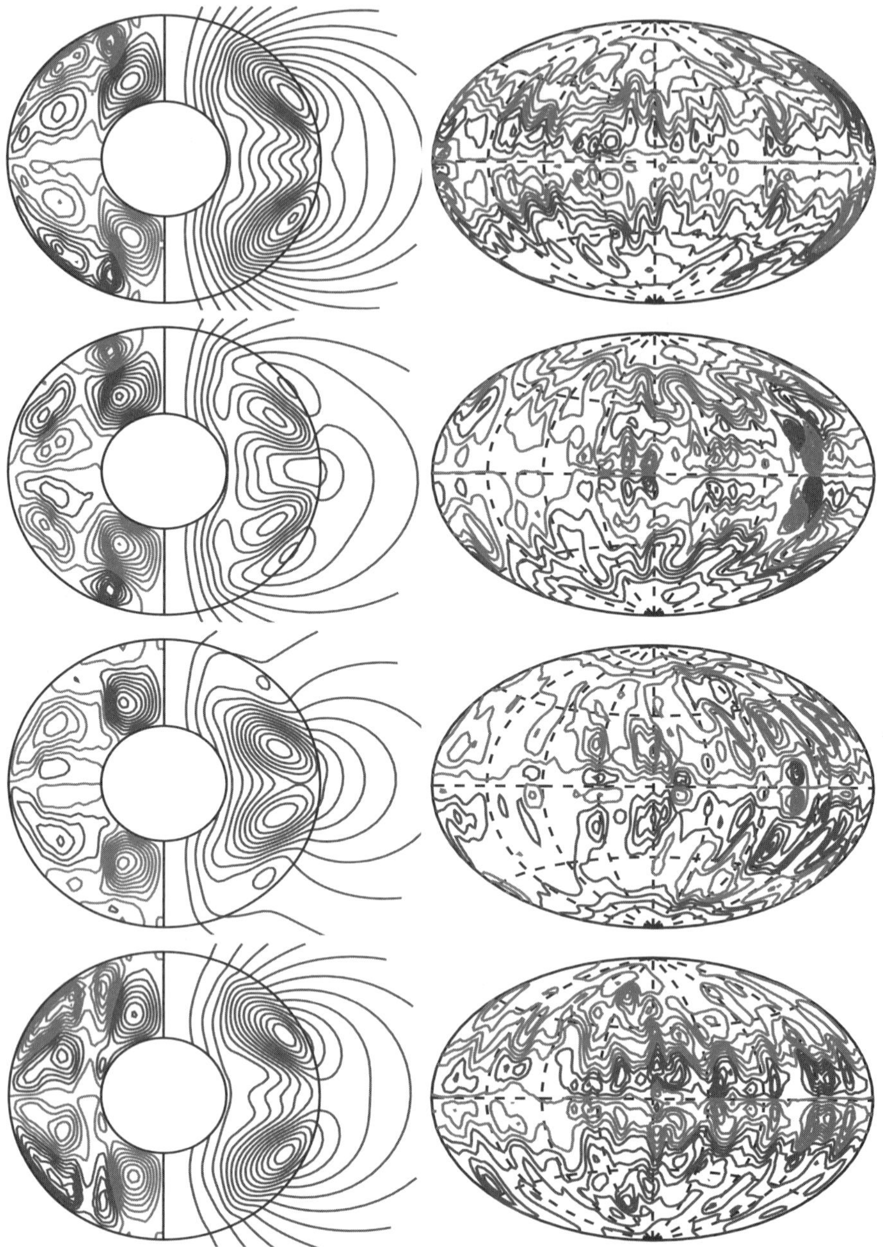

Fig. 6.13. Oscillating dipolar dynamo for $\tau = 5 \cdot 10^3, R = 6 \cdot 10^5, P = P_m = 5$. The time sequences of plots (from top to bottom with $\Delta t = 0.08$) show lines of constant azimuthally averaged magnetic field component $\bar{B}_\varphi$ in the left halves of the circles. In the right halves meridional field lines are shown. The plots on the right show lines of constant radial component B_r at $r = r_i + 1.3d$ where r_i is the inner radius and d is the thickness of the shell. The plots cover roughly one period of the oscillations of the chaotic dynamo convection state.

duction of the critical Rayleigh number for onset of convection by an imposed magnetic field which was found by Chandrasekar [47] for a rotating plane layer and by Fearn [48] for a rotating sphere. A reduction of the critical Rayleigh number for caused by the dynamo process has not yet been found.

Another interesting feature of convection-driven dynamos in rotating fluid shells is the fact that the dynamo process may occur in an oscillatory fashion outside the cylindrical surface touching the inner core at the equator, whereas the field remains stationary in the polar regions inside this surface. Figure 6.13 shows an example for this property obtained in recent dynamo simulations of Simitev and Busse [49]. Because the magnetic field observed at the Earth's surface mainly reflects properties of the dynamo process in the high-latitude regions of the outer core, it is thus conceivable that the geodynamo includes an oscillatory process that is not evident from paleomagnetic measurement. Only occasionally it may lead to a reversal of the entire geomagnetic field.

6.5 Concluding Remarks

The framework of this brief survey of the role played by convection processes in geophysics and astrophysics is much too restricted to do justice to this flourishing and growing field of research. The few examples discussed in the preceding sections should simply demonstrate that the visions of Bénard and his coworkers about the relevance of simple laboratory convection experiments to a broad range of geophysical and astrophysical phenomena are still with us today. Laboratory experiments and simple theoretical models continue to stimulate much of today's fluid dynamical studies in geophysics and astrophysics in spite of the increasing role played by large-scale computer simulations. The solar dynamo, the unusual dynamos of Neptune and Uranus and last but not least the geodynamo are still not fully understood. In the Earth's core the interplay of thermal buoyancy and of chemical buoyancy generated by light elements left in solution by the crystallizing inner core has received little attention. There are thus many reasons that the investigations of convection processes will occupy geophysicists and astrophysicists for a long time in the future.

References

1. L. Prandtl, *The Essentials of Fluid Dynamics*, Blackie, London (1952).
2. J.A.Whitehead and B. Parsons, Observations of convection at Rayleigh numbers up to 760000 in a fluid with large Prandtl number, *Geophys. Astrophys. Fluid Dyn.* **9**, 201–217 (1978).
3. J.R.A. Pearson, On convection cells induced by surface tension, *J. Fluid Mech.* **4**, 489–500 (1958).
4. R.C. DiPrima and H.L. Swinney, Instabilities and transition in the flow between concentric rotating cylinders, in *Hydrodynamic Instabilities and Transition to Turbulence*, 139–180, ed. by H.L. Swinney and J.P. Gollub, Springer-Verlag, New York (1981).

5. D. Avsec, Sur les formes ondulées des tourbillons en bandes longitudinales, *Compt. Rend Acad. Sci.* **204**, 167–169 (1937).

6. H. Bénard and D. Avsec, Travaux récent sur les tourbillons en bandes: Applications à l'Astrophysique et à la Météorologie, *J. de Physique et le Radium* **9**, 486–500 (1938).

7. H. Bénard, Les tourbillons cellulaires dans une nappe liquide transportant de la chaleur par convection en régime permanent, *Ann. Chim. Phys.* **7** Ser **23**, 62 (1901).

8. F.H. Busse and J.A. Whitehead, Instabilities of convection rolls in a high Prandtl number fluid, *J. Fluid Mech.* **47**, 305–320 (1971).

9. J.A. Whitehead, The propagation of dislocations in Rayleigh–Bénard rolls and bimodal flow, *J. Fluid Mech.* **75**, 715–720 (1976).

10. B.W. Atkinson and J. Wo Zhang, Mesoscale shallow convection in the atmosphere, *Rev. Geophysics* **34**, 403–431 (1996).

11. A. Graham, Shear patterns in an unstable layer of air, *Phil. Trans. Roy. Soc.* **A232**, 285–296 (1933).

12. H. Tippelskirch, Über Konvektionszellen, insbesondere in flüssigem Schwefel, *Beitr. Phys. Atmos.* **29**, 37–54 (1956).

13. F.H. Busse, The stability of finite amplitude cellular convection and its relation to an extremum principle, *J. Fluid Mech.* **30**, 625–649 (1967).

14. M. Assenheimer and V. Steinberg, Observations of coexisting up- and downflow hexagons in Boussinesq Rayleigh–Bénard convection, *Phys. Rev. Lett.* **76**, 756–759 (1996).

15. R.M. Clever and F.H. Busse, Hexagonal convection cells under conditions of vertical symmetry, *Phys. Rev. E* **53**, R2037–R2040 (1996).

16. A. Wegener, Die Entstehung der Kontinente, Petermanns Mitt. 1912, 185–195, 253–256, 305–309 (1912).

17. A. Holmes, Radioactivity and earth movements, *Geol. Soc. Glasgow Trans.* **18**, 559–606 (1931).

18. F.H. Busse and N. Riahi, Patterns of convection in spherical shells II, *J. Fluid Mech.* **47**, 305–320 (1971).

19. F.H. Busse, Patterns of convection in spherical shells, *J. Fluid Mech.* **72**, 67–85 (1975).

20. J.-P. Montagner, Can seismology tell us anything about convection in the mantle?, *Rev. Geophys.* **32**, 115–137 (1994).

21. G. Masters, S. Johnson, G. Laske, and H. Bolton, A shear-velocity model of the mantle, *Phil. Trans. R. Soc. Lond.* **A354**, 1385–1411 (1996).

22. F.H. Busse, Quadrupole convection in the lower mantle? *Geophys. Res. Lett.* **10**, 285–288 (1983).

23. W.R. Peltier, Editor, *Mantle Convection, Plate Tectonics and Global Dynamics*, Gordon & Breach, New York (1989).

24. G. Schubert, D.L. Turcotte, and P. Olson, *Mantle Convection in the Earth and Planets*, Cambridge University Press (2001).

25. J. Janssen, *Ann. Observ. Astron. Phys. Paris, sis á Meudon*, t.1, p. 103 (1896).

26. F.H. Busse, Differential rotation in stellar convection zones, *Astrophys. J.* **159**, 629–639 (1970).

27. G.W. Simon and N.O. Weiss, Supergranules and the hydrogen convection zone, *Z. Astrophys.* **69**, 435–450 (1968).

28. F.H. Busse, On Howard's upper bound for heat transport by turbulent convection, *J. Fluid Mech.* **37**, 457–477 (1969).

29. F.H. Busse, Thermal instabilities in rapidly rotating systems, *J. Fluid Mech.* **44**, 441–460 (1970).

30. F.H. Busse and L.L. Hood, Differential rotation driven by convection in a rotating annulus, *Geophys. Astrophys. Fluid Dyn.* **21**, 59–74 (1982).

31. F.H.Busse, Asymptotic theory of convection in a rotating, cylindrical annulus, *J. Fluid Mech.* **173**, 545–556 (1986).

32. A.C. Or and F.H. Busse, Convection in a rotating cylindrical annulus. Part 2. Transitions to asymmetric and vacillating flow, *J. Fluid Mech.* **174**, 313–326 (1987).

33. E. Grote and F.H. Busse, Dynamics of convection and dynamos in rotating spherical fluid shells, *Fluid Dyn. Res.* **28**, 349-368 (2001).

34. J. Larmor, How could a rotating body such as the sun become a magnet? *Brit. Ass. Advan. Sci. Rep.*, 159–160 (1919).

35. T.G. Cowling, The magnetic field of sunspots, *Monthly Not. Roy. Astr. Soc.* **94**, 39–48 (1934).

36. G.E. Backus, A class of self sustaining dissipative spherical dynamos, *Ann. Phys.* **4**, 372–447 (1958).

37. A. Herzenberg, Geomagnetic dynamos, *Phil. Trans. Roy. Soc. London, Ser. A* **250**, 543–585 (1958).

38. S. Childress and A.M. Soward, Convection-driven hydromagnetic dynamo, *Phys. Rev. Lett.* **29**, 837–839 (1972).

39. A.M. Soward, A convection driven dynamo. I. The weak field case, *Phil. Trans. Roy. Soc. London, Ser. A* **275**, 611–645 (1974).

40. F.H. Busse, Generation of magnetic fields by convection, *J. Fluid Mech.* **57**, 529–544 (1973).

41. F.H. Busse, A model of the geodynamo, *Geophys. J. R. Astr. Soc.* **42**, 437–459 (1975).

42. G.A. Glatzmaier, Numerical simulations of stellar convective dynamos. I. The model and the method, *J. Comp. Phys.* **55**, 461–484 (1984).

43. F.H. Busse, Homogeneous dynamos in planetary cores and in the laboratory, *Ann. Rev. Fluid Mech.* **32**, 383–408 (2000).

44. E. Dormy, J.-P. Valet, and V. Courtillot, Numerical models of the geodynamo and observational constraints, G^3 *(Geochem. Geophys. Geosystems)* **1**, (2000).

45. D.R. Fearn, Hydromagnetic flow in planetary cores, *Rep. Prog. Phys.* **61**, 175–235 (1998).

46. P.H. Roberts and G.A. Glatzmaier, Geodynamo theory and simulations, *Rev. Mod. Phys.* **72**, 1081–1123 (2000).

47. S. Chandrasekar, *Hydrodynamic and Hydromagnetic Stability*, Clarendon Oxford (1961).

48. D.R. Fearn, Thermally driven hydromagnetic convection in a rapidly rotating sphere, *Proc. Roy. Soc. Lond.* **A369**, 227–242 (1979).

49. R. Simitev and F.H. Busse, Visible and nearly invisible oscillatory convection driven dynamos, to be submitted to *Geophys. Res. Lett.* (2003).

50. F.H. Busse, and N. Riahi, Patterns of convection in spherical shells II, *J. Fluid Mech.* **123**, 283–302 (1982).

Part III

Bénard–Marangoni Cellular Structures

7 Bénard Layers, Overstability, and Waves

Manuel G. Velarde

[1] Instituto Pluridisciplinar, Universidad Complutense de Madrid, Paseo Juan XXIII,
n. 1, 28040–Madrid, Spain

[2]

International Center for Mechanical Sciences (CISM),
Palazzo del Torso, Piazza Garibaldi, 33100–Udine, Italy
velarde@fluidos.pluri.ucm.es

Provided here are predictions concerning overstability and the onset of oscillatory
motions leading to transverse capillary–gravity waves or longitudinal (dilational)
waves due to the Marangoni effect in a Bénard layer heated from the air side
above or subject to the ad/absorption of a relatively light component hence
leading to an apparently stably stratified state.

7.1 Introduction

The onset of Bénard convection and the evolution of Bénard cells when heating
a shallow from the bottom of the liquid side below, driven by surface tension
gradients and the Marangoni effect, have been intensively studied in the past
decade, at least for moderate values of the thermal gradient. Two recent monographs [1, 2] and a review paper [3] provide detailed information about the
present understanding of the problem. It remains, however, to study the evolution of the flow pattern when the thermal gradient takes high values, and hence
the corresponding Marangoni (Péclet) number grows large. This is the realm of
dissipative interfacial turbulence where inertia, although needed, plays a much
less significant role than in Reynolds–Kolmogorov turbulence (which is a nonlinear problem where dissipation becomes significant only at the end of the energy
transfer cascade when little vortices are killed by viscosity). Particularly interesting for dissipative interfacial turbulence is the case of liquids with high values of
the Prandtl number where the velocity field is slaved (vanishing usual Reynolds
number) by the temperature field and where the tangential boundary condition
with the Marangoni effect provides the velocity scale and the energy input to
the system. I wonder if, due to the fact that dissipation plays such a significant
role, a cascade from short to large scales appears as the latter dissipate less.

Another interesting problem, much less studied than steady cellular Bénard
convection is the onset of overstability eventually leading to wave motion driven
by the same Marangoni effect but when the liquid layer is heated from above. Although recent monographs and various papers touch upon the problem, our understanding is still fragmentary [2, 4, 5, 6, 7, 8]. For the commemorative purposes
of the Bénard centennial and to attract the interest of experimentalists, I recount
here a few salient features of this problem dealing, in the simplest possible and
physically appealing way, with threshold values and instability mechanisms.

When the equilibrium surface of a liquid layer open to ambient air is disturbed, the interface behaves as does a membrane; it gets deformed and consequently forces tending to return it to the original equilibrium state appear in the liquid. Capillary forces (surface tension) tend to reduce the increased surface and, if the liquid is in a gravitational field, gravitational forces tend to return the interface to its original equipotential level shape. However, because of inertia the liquid particles may overshoot their original equilibrium position (overstability). As a consequence transverse waves may appear at the air–liquid interface (capillary–gravity waves). Viscosity tends to damp these waves. For a frequency of oscillation Ω and kinematic viscosity ν, the viscous penetration length of the wave is of order $(\Omega/\nu)^{1/2}$.

Generally, for high enough frequencies, $\Omega \gg \nu/d^2$, the (viscous) wave penetration depth is rather small with respect to the layer depth d, and viscosity can be neglected in the bulk away from the interface. Accordingly, capillary–gravity waves are directly controlled by stresses orthogonal to the interface and they can be described, to a first –albeit significant– approximation using inviscid liquid theory. Their dispersion relation is expected to depend on both surface tension and gravitational acceleration, and indeed density. Note, however, that if a spontaneous fluctuation misaligns the surface deformation relative to the isotherms, the Marangoni effect may help to bring the surface back to its equilibrium–level position or, otherwise, depending on the sign of the thermal gradient, may sustain the disturbance.

There is another possible wave disturbance, with motion mostly longitudinal along the interface, and hence directly controlled by the tangential stresses as discussed by Lucassen [9]. However, he only considered damped motions and said nothing about a possible instability allowing the compression–expansion like waves to be sustained. Its existence is not surprising if surfactants are present and one takes the analogy between a monolayer–covered surface and an elastic membrane. Compression–expansion motions are possible eventually leading to interfacial waves whose dispersion relation is not related to surface tension but rather to its variation along the interface. These waves are not affected by the gravitational field and the associated motion is also expected to penetrate little in the bulk of the liquid. The variation of the surface tension along the interface brings into play the Marangoni effect and hence whether there are surfactants present is immaterial provided this effect is present. Such could be the case even in incompressible liquids if there is heat transfer along or across the interface.

I show how, past an instability threshold, wave motion can be sustained and I provide the corresponding oscillation frequency and critical Marangoni numbers in terms of Prandtl, or Schmidt, Bond, Galileo, and capillary numbers (defined below). For the sake of completeness, I first consider the case of heat transfer leading to capillary–gravity waves. Then I proceed to a study of both transverse and longitudinal waves when surfactants are present. Both heat and surfactant transfer lead, for transverse motions, to exactly the same capillary–gravity waves. The same procedure applies to longitudinal motions but I discuss only the case of surfactants. On the other hand, to use a physically appealing picture, I identify the harmonic oscillator underlying either transverse or longitudinal motions, thus making transparent the role of the Marangoni effect both in the elastic spring constant and in the energy supply term needed to overcome damping in the system.

7.2 Heat Transfer and Capillary–Gravity Waves (Laplace–Kelvin Waves)

At an air–liquid interface, hence at the open surface of a heated liquid layer, as in Bénard's experiments, flow disturbances obey the Navier–Stokes equations, that in the simplest (x, z) two-dimensional (2-D) geometry reduce to

$$\frac{\partial w}{\partial t} = -\frac{1}{\rho}\frac{\partial p}{\partial z} + \nu\left(\frac{\partial^2 w}{\partial x^2} + \frac{\partial^2 w}{\partial z^2}\right), \tag{7.1}$$

where, for simplicity, we restrict consideration to an infinitely extended horizontal liquid layer; t and z denote time and the vertical coordinate, and w, ρ and p are vertical velocity, density, and pressure, respectively. As w is the liquid velocity, the kinematic condition, $w = \partial\xi/\partial t$, at every time instant t provides a direct relationship between the velocity of a point at the geometrical surface and that of a material liquid point. $\xi(x, t)$ denotes the deformation of the interface. Neglecting the mechanical influence of the ambient air (the *dynamic* shear viscosity of air is, generally, much lower than for a liquid), at the interface the normal and tangential components of the stress tensor are

$$p - \rho g\,\xi + \sigma\frac{\partial^2\xi}{\partial x^2} = 2\,\eta\,\frac{\partial w}{\partial z} \tag{7.2}$$

and

$$\eta\left(\frac{\partial u}{\partial z} + \frac{\partial w}{\partial x}\right) = \frac{\partial\sigma}{\partial T}\left(\frac{\partial T}{\partial x} - \beta\frac{\partial\xi}{\partial x}\right), \tag{7.3}$$

where u is the corresponding horizontal velocity component, g is the gravitational acceleration, σ is the surface tension, T denotes temperature, and β is the temperature gradient imposed across the liquid layer (β is here taken positive when the layer is heated from the liquid side below); $\eta = \rho\nu$ is the dynamic shear viscosity. Equation (7.2) shows that I do take gravity into account; otherwise the capillary length (defined below) diverges to infinity which is unphysical and, moreover, does not suit my purposes here. However, I disregard buoyancy in the bulk.

In the high-frequency limit, the bottom of the liquid layer can be considered at "practical" infinity from the interface. Then the liquid depth d becomes useless and a suitable scale is the capillary length obtained when the static Bond number $Bo = \rho g l^2/\sigma$, is set equal to unity. Then using Equations (7.1) to (7.3), the heat Fourier diffusion equation, and the kinematic condition, one obtains

$$\frac{d^2\xi}{dt^2} + k^2\,K\,\frac{d\xi}{dt} + L\,\xi = 0, \tag{7.4}$$

where

$$K = \left[4\nu - \frac{k\beta}{\rho}\frac{\partial\sigma}{\partial T}\left(\frac{\chi}{2\Omega^3}\right)^{1/2}\right],$$

$$L = \left\{gk + \frac{\sigma k^3}{\rho}\left[1 + \frac{\beta}{\sigma}\frac{\partial\sigma}{\partial T}\left(\frac{\chi}{2\Omega}\right)^{1/2}\right]\right\},$$

and I have assumed that

$$\xi = -\frac{A}{\Omega}\cos\left(kx + \Omega t\right), \tag{7.5}$$

and similar Fourier normal mode expressions for all other disturbances; χ is the thermal diffusivity.

Equation (7.4) is the harmonic oscillator equation obeyed by a point at the air–liquid interface. The damping coefficient may be positive, negative, or vanishing according to the sign and values given to β for given $\partial\sigma/\partial T$. Using the capillary length as unit, $l = (\sigma/\rho g)^{1/2}$, Equation (7.4) takes on a dimensionless, universal form

$$\frac{d^2\zeta}{d\tau^2} + \tilde{K}\frac{d\zeta}{d\tau} + \tilde{L}\,\zeta = 0, \tag{7.6}$$

where

$$\tilde{K} = \left[4a^2 + \frac{M\,a^3}{(2\,\mathrm{Pr}^3\,\omega^3)^{1/2}}\right]$$

and

$$\tilde{L} = \left[\frac{C\,a\left(Bo + a^2\right)}{\mathrm{Pr}} - \frac{M}{a\left(2\,\mathrm{Pr}^3\,\omega^3\right)^{1/2}}\right]$$

with $\zeta = \xi/l$, $\tau = t\nu/l^2$, $a = lk$, and $\omega = \Omega/\nu k^2$. Three dimensionless groups appear: the Marangoni number $M = -\left(\partial\sigma/\partial T\right)\beta l^2/\eta\chi$, the Prandtl number $Pr = \nu/\chi$ and the capillary number $C = \sigma l/\eta\chi$. The Bond number explicitly shows the relative influence of surface tension to gravity. Further simplification of Equation (7.6) can be achieved by redundantly restricting consideration to the above–invoked high–frequency limit. We see that in order to have vanishing damping it suffices to set $M = 0\left(\omega^{3/2}\right)$, with $Pr = 0\,(1)$. We can also set $C = 0\left(\omega^2\right)$, which is a reasonable assumption provided we are far from a critical point or if the liquid layer is not too shallow. Then the second term in the spring constant, the coefficient of ζ, can be neglected and Equation (7.6) reduces to

$$\frac{d^2\zeta}{d\tau^2} + \left[4a^2 + \frac{M\,a^3}{(2\,\mathrm{Pr}^3\,\omega^3)^{1/2}}\right]\frac{d\zeta}{d\tau} + \left[\frac{C\,a\left(Bo + a^2\right)}{\mathrm{Pr}}\right]\zeta = 0, \tag{7.7}$$

which is the simplest (high-frequency) harmonic oscillator approximation to the oscillatory transverse interfacial motion of the open surface in a liquid layer. It contains, however, all the relevant physics. Indeed, the damping coefficient vanishes when I set

$$M = -\frac{4}{a}\left(2\,\overset{3}{\mathrm{Pr}}\,\omega^3\right)^{1/2}, \tag{7.8}$$

thus allowing a "free" oscillation of (dimensionless) frequency given by

$$\omega^2 = \frac{C\,a(Bo + a^2)}{Pr} = \tilde{G}\,a\left(1 + \frac{a^2}{Bo}\right) \tag{7.9}$$

or else, in dimensional form,

$$\Omega^2 = gk + \frac{\sigma k^3}{\rho}, \tag{7.10}$$

which is the dispersion relation for capillary–gravity waves in an infinitely deep liquid layer (Laplace–Kelvin law). Note that $C/Pr = \tilde{G}/Bo$, $Pr\,\tilde{G} = G$, thus defining G and $\tilde{G}$. The dispersion relation, for a layer of finite depth, is $\omega^2 = \tilde{G}\,a\left(1 + a^2/Bo\right)\tanh a$.

It does appear that for an air–liquid interface the oscillatory instability is to be expected for negative values of the Marangoni number only. For standard liquids $(\partial\sigma/\partial T < 0)$, this means that the heating is from the ambient air above or the layer is cooled from the liquid bottom below. The minimum Marangoni number needed to sustain the oscillatory motion is

$$Ma_c = -7.93\,(CPr)^{3/4} \tag{7.11}$$

with a frequency

$$\omega_c^2 = 6\sqrt{5}C\Big/Pr \tag{7.12}$$

and a wave number

$$a_c = 1/\sqrt{5}. \tag{7.13}$$

These results have also been obtained using a formally different approach [10] (see also [4] for a full list of references).

It seems interesting to see how temperature and surface deformation appear in the wave motion. To do this, I need to recall the explicit forms of the Fourier normal modes. Besides Equation (7.5) we have

$$u = \left(Ae^{kz} + Be^{mz}\right)e^{ikx+\lambda t} \tag{7.14}$$

and

$$w = -i\left(A\,e^{kz} + \frac{kB}{m}e^{mz}\right)e^{ikx+\lambda t} \tag{7.15}$$

with $m^2 = \left(k^2 + \Omega/\nu\right)$; λ is the complex time constant or growth rate. Note that I could have taken $e^{\lambda t}$ or $e^{-\lambda t}$, according to convention. Then $Re\lambda > 0$ (alternatively, $Re\lambda < 0$) implies instability, $Re\lambda = 0$ denotes a neutrally stable case, and $Im\lambda \equiv \Omega$ denotes the frequency of overstable oscillatory motions. On the other hand, $Re\lambda < 0$ (alternatively, $Re\lambda > 0$) implies (linear or local) stability. Linear stability analyses provide sufficient conditions for instability or necessary conditions for stability. A system may be linearly stable but unstable to finite amplitude disturbances (a case not considered here). The equation

$Im\lambda = 0$ provides the dispersion relation between frequency and corresponding wavenumber (or wavelength).

The complete expression of the mode (7.5) is

$$\xi = -\frac{i}{\lambda}\left(A + \frac{kB}{m}\right)e^{ikx+\lambda t}, \tag{7.16}$$

according to the above indicated kinematic boundary condition at $z = 0$. Equation (7.5) provides the real part of (7.16). Then in the earlier-invoked high-frequency approximation the boundary condition involving the Marangoni effect (7.3) yields

$$B = \left(\frac{2}{\omega}\right)^{1/2}(3+i)A + O(\omega^{-1/2}) \tag{7.17}$$

that, as expected, shows that for transverse oscillations the rotational part is negligible in comparison with the potential one in the Fourier normal mode solutions.

From the velocity field, (7.14) and (7.15), it follows that the temperature disturbance along the interface is

$$T(0) - \beta\xi = \frac{A\beta}{\Omega^{3/2}}\chi^{1/2}k\,\sin(kx + \Omega t + \frac{\pi}{4}). \tag{7.18}$$

Comparison of (7.18) and (7.16) shows that temperature and interfacial deformation are such that the extrema of the former disturbance do not sit on those of the latter, that is, on the crests and troughs of the wave for the hottest and coldest spots, respectively (Figure 7.1). Then wave propagation can be enhanced or suppressed by the Marangoni effect, according to the sign of the temperature gradient. When for a standard liquid, the heating is from the air side (β negative) the Marangoni effect brings liquid from the hot point, located, for example, at $x = 5\pi/4$, to the cold point at $x = \pi/4$. Such flow favors the motion assumed for the wave, right to left. In the opposite case when for the same standard liquid we heat from the liquid side or cool it from the air above (β positive) the hot point is, for example, at $x = \pi/4$, whereas the cold point is at $x = 5\pi/4$.

Now the Marangoni flow rather pushes the interfacial disturbance to the right, opposes the wave propagation and eventually suppresses it. Thus we have seen that transverse vibrations, and hence capillary–gravity waves, at the open surface of a liquid can be excited, and eventually sustained, by the Marangoni effect. Note that, by providing a sufficient condition for instability, an approximate linear theory (done here for high–frequency motions) may very well overestimate the experimental value of the threshold. In the description presented above I have used the capillary length as a natural scale and hence the liquid layer was tacitly taken of infinite depth but not quite, only in the sense that viscous effects were appreciable only very near the open surface of the liquid. If the bottom of the layer affects the open surface it is expected to play a stabilizing role hence lowering the above given threshold value. The asymptotics used is also very limited in scope. Thus, to have a complete picture of overstability, with better quantitative estimates of its characteristic quantities, a two–layer system with the full

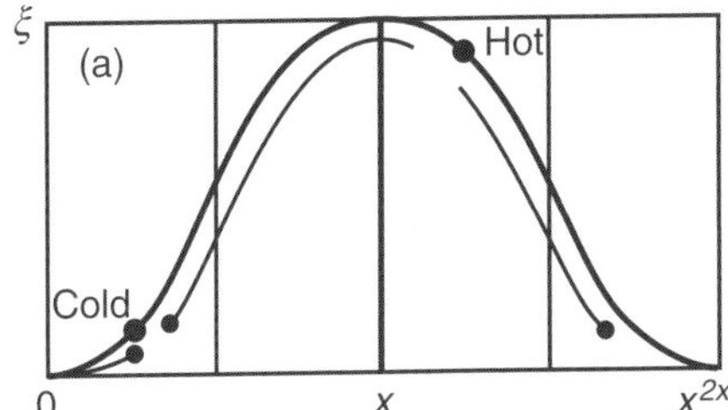

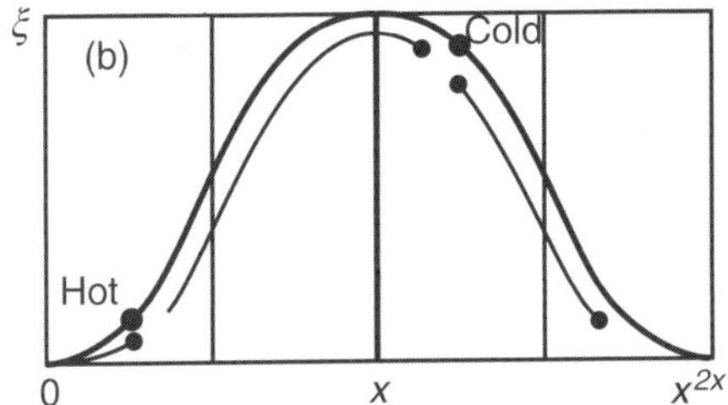

Fig. 7.1. Transverse waves. Schematic, and exaggerated, sketch of the phase shift between temperature (7.18) disturbances (hot and cold spots only) and surface deformation ξ (7.11), in a capillary–gravity wave moving right to left. (a) The liquid layer is heated from the ambient air above ($\beta < 0$); (b) the heating is done from the liquid side below ($\beta > 0$). For standard liquids the Marangoni effect acts according to the arrows (flow from hot to cold regions), and hence favors flow disturbance in the former case whereas it tends to suppress it in the latter.

thermohydrodynamic equations needs to be considered with full description of boundaries and boundary layers at the solid bottom and at the open surface as well. Noteworthy is the fact that when two liquid layers are considered in full, waves may appear for either sign of the Marangoni number, depending on the ratio of (kinematic) viscosities and heat diffusivities.

7.3 Surfactants and Interfacial Waves

7.3.1 Disturbance Equations

To make the presentation almost self-contained and to clarify the limitations of the role of surfactant transport here considered, let me recall the evolution equations obeyed by flow disturbances and surfactant concentration [11]. Again, as in the heat transfer problem, for simplicity I consider a horizontally infinitely extended two–dimensional $(x,\ z)$ liquid layer with infinite depth. The motionless undisturbed surface is located at $z = 0$. The equations that small disturbances upon the quiescent state obey are

$$\nabla \cdot \boldsymbol{v} = 0, \tag{7.19}$$

$$\rho \frac{\partial \boldsymbol{v}}{\partial t} = -\nabla p + \eta \nabla^2 \boldsymbol{v}, \tag{7.20}$$

$$\frac{\partial c}{\partial t} - \beta^c w = D \nabla^2 c, \tag{7.21}$$

where $\boldsymbol{v} = (u,\ w)$, p is the velocity pressure, c the surfactant concentration, ρ the density of the liquid, D is the mass diffusivity, and $\beta^c = (\partial c/\partial z)_0$ denotes the surfactant gradient at the quiescent state; all other symbols are as earlier defined. The disturbances obey the following (linearized) boundary conditions at the deformable surface [12, 13]

$$\frac{\partial \zeta}{\partial t} = w, \qquad (7.22)$$

$$-T_0 \nabla_\alpha^2 \zeta + g\rho\zeta - p - 2\eta \frac{\partial w}{\partial z} = 0, \qquad (7.23)$$

$$\left(\frac{\partial T}{\partial \Gamma}\right)_0 \nabla_\alpha \gamma - \eta \left(\nabla_\alpha w + \frac{\partial u}{\partial z}\right) = 0, \qquad (7.24)$$

$$\frac{\partial \gamma}{\partial t} + \Gamma_0 \nabla_\alpha * u_\alpha - D_\alpha \nabla_\alpha^2 \gamma + D \frac{\partial c}{\partial z} = 0, \qquad (7.25)$$

$$\gamma = k^1 \left(c - \beta^c \zeta\right)_\alpha, \qquad (7.26)$$

where ζ is the surface deviation from the $z = 0$ level, Γ is the excess surfactant concentration at the surface, the subscript 0 indicates a value in a reference state, and γ is the disturbance on Γ_0; note that here T is the surface tension, and the subscript α accounts for either a value taken on the surface or a derivative along the surface.

New units allow us to rescale the quantities in the equations. As in the preceding analysis the capillary length $l = (T_0/\rho g)^{1/2}$ is chosen as the length scale; ν/l, l^2/ν, $\nu^2\rho/l^2$, $\beta^c l$, and Γ_0 are used as units for velocity, time, pressure, surfactant concentration and *excess* surface concentration, respectively. Thus, Equations (7.15) to (7.22) become

$$\nabla \cdot \boldsymbol{v} = 0, \qquad (7.27)$$

$$\frac{\partial \boldsymbol{v}}{\partial t} = -\nabla p + \nabla^2 \boldsymbol{v}, \qquad (7.28)$$

$$\frac{\partial c}{\partial t} - w = S^{-1} \nabla^2 c \qquad (7.29)$$

with boundary conditions at $z = \zeta$,

$$\frac{\partial \zeta}{\partial t} = w, \qquad (7.30)$$

$$-\frac{1}{SC} \nabla_\alpha^2 \zeta + \frac{Bo}{SC} \zeta - p - 2\frac{\partial w}{\partial z} = 0, \qquad (7.31)$$

$$\frac{HE}{SH_z} \nabla_\alpha \gamma + \left(\nabla_\alpha \gamma + \frac{\partial u}{\partial z}\right) = 0, \qquad (7.32)$$

$$HS \left(\frac{\partial \gamma}{\partial t} + \nabla_\alpha * u_\alpha - S_\alpha^{-1} \nabla_\alpha^{-2} \gamma\right) + \frac{\partial c}{\partial z} = 0, \qquad (7.33)$$

$$\gamma = \frac{H_z}{H} \left(c - \zeta\right)_\alpha, \qquad (7.34)$$

where expecting no confusion for the reader I have used the same notation for t, v, p, c, γ, and ζ in these dimensionless equations, as in the dimensional case. The following dimensionless parameters have been introduced: $S = v/D$ is the Schmidt number; $S_\alpha = v/D_\alpha$ is the surface Schmidt number; $C = lT_0/\eta D$, now is the capillary number; $Bo = \rho g l^2/T_0$ is the Bond number; $E = -\left(\partial c/\partial \Gamma\right)_0 k^l \beta l^2 \left(\eta D\right)$ is the surfactant (elasticity) Marangoni number; $H = \Gamma/\left(\beta^c l^2\right)$ is the surface excess surfactant number; and $H_z = k^l/l$ denotes the Langmuir adsorption number [12, 13, 14]. Then Equations (7.15)to (7.22) have solutions like (7.14), (7.15), and so on,

$$w = \left(Ae^{az} + Be^{mz}\right) e^{iax+\lambda t}, \tag{7.35}$$

$$u = \left(iAe^{az} + imBe^{mz}\right) e^{iax+\lambda t}, \tag{7.36}$$

$$p = -\left(\frac{\lambda Ae^{az}}{a}\right) e^{iax+\lambda t}, \tag{7.37}$$

$$c = \left(\frac{A}{\lambda}e^{az} + \frac{SB}{\lambda\left(S-l\right)}e^{mz} + Fe^{qz}\right) e^{iax+\lambda t}, \tag{7.38}$$

where here a denotes the Fourier mode with $m^2 = a^2 + \lambda$, $q^2 = a^2 + S\lambda$, and $Im\lambda = \omega$.

7.3.2 Transverse Capillary-Gravity Waves (Laplace–Kelvin Waves)

Assuming, for further simplicity, that adsorption and surface accumulation have a negligible effect on the transverse surface waves, the above-posed problem reduces to

$$\frac{Ea^2}{S}\left(c-\zeta\right) + \left(\frac{\partial^2}{\partial z^2} + a^2\right) w = 0, \tag{7.39}$$

$$\frac{\partial c}{\partial z} = 0. \tag{7.40}$$

As for the heat transfer problem, Equation (7.16) gives the evolution equation of the liquid layer. It is valid in the volume as well as at the surface, a part of the liquid. On the other hand, there is a kinematic relation (7.18) on the surface. Then at $z = \zeta$, we have

$$\frac{\partial^2 \zeta}{\partial t^2} = -\frac{\partial p}{\partial z} + \nabla^2 w. \tag{7.41}$$

Noting that $\partial p/\partial z = ap$, equation (7.41) becomes

$$\frac{\partial^2 \zeta}{\partial t^2} + \frac{Bo + a^2}{SC}a\zeta = -2a\frac{\partial w}{\partial z} - 2a^2 w - \frac{Ea^2}{S}\left(c-\zeta\right). \tag{7.42}$$

To proceed further, I need the coefficients A, B, and F. We have

$$B = \frac{\lambda f\left(Bo\right) + 2a}{\lambda f\left(Bo\right) - \lambda/a - 2p}A \tag{7.43}$$

with

$$f\left(Bo\right) = \frac{Bo + a^2}{SC\lambda^2} + \frac{1}{a}.$$
(7.44)

For transverse waves the Laplace–Kelvin equation $f\left(Bo\right) = 0$ (see also equation (7.9) or (7.10)) defines the dispersion relation for capillary–gravity waves, and as the interfacial disturbance penetrates little in the liquid, $\omega >> a^2$. Then from Equation (7.31) it follows that

$$B \approx -\frac{2a^2}{\lambda}A << A.$$
(7.45)

Clearly, as for heat transfer, the potential part is the significant ingredient in the velocity field, which is consistent with the fact that transverse waves do exist in inviscid fluids. Taking advantage of this simplification, Equation (7.42) becomes

$$\frac{d^2\zeta}{dt^2} + \frac{Bo + a^2}{SC}\,a\zeta = -4a^2\frac{dz}{dt} - \frac{Ea^2}{S}\left(c - \zeta\right).$$
(7.46)

The coefficient F can be obtained from Equation (7.38):

$$F = -\frac{1}{q}\left(\frac{a}{\lambda}A + \frac{mS}{\lambda\left(S - 1\right)}B\right) \approx -\frac{a}{q\lambda}A,$$
(7.47)

which leads to

$$c - \zeta = \left(F + \frac{B}{\lambda\left(S - l\right)}\right)e^{iax + \lambda t} \approx -\frac{aA}{q\lambda}e^{iax + \lambda t} \approx a\frac{d\zeta/dt - \omega\zeta}{\omega\sqrt{2S\omega}}.$$
(7.48)

Thus, Equation (7.46) becomes

$$\frac{d^2\zeta}{dt^2} + \left(4a^2 + \frac{Ea^3}{S\omega\sqrt{2S\omega}}\right)\frac{d\zeta}{dt} + \left(aC\frac{Bo + a^2}{S} - \frac{Ea^3}{S\sqrt{2S\omega}}\right)\zeta = 0,$$
(7.49)

and assuming that $E = 0\left(\omega^{3/2}\right)$, $C = 0\left(\omega^2\right)$, and $S = 0\left(1\right)$, it follows that

$$\frac{C\,a(B + a^2)}{S} >> \frac{E\,a^3}{S\,\sqrt{2S\omega}}.$$

Therefore, Equation (7.49) can be reduced to

$$\frac{d^2\zeta}{dt^2} + \left(4a^2 + \frac{Ea^3}{S\omega\sqrt{2S\omega}}\right)\frac{d\zeta}{dt} + \frac{Bo + a^2}{SC}\,a\zeta = 0.$$
(7.50)

Equation (7.50) describes the oscillatory motion of the surface and takes the form of a harmonic oscillator. In the absence of dissipation and the Marangoni effect, the equation appears as an ideal oscillator as potential motion is indeed the main character of a transverse wave. At the neutral state,

$$E = -4\frac{(2S^3\omega^3)^{1/2}}{a},$$
(7.51)

quite the same as equation (7.8) for the heat transfer problem, save the change P for S.

The kinetic energy dissipated by viscosity and that produced by surface tension work (Marangoni effect) just compensate each other, sustaining the oscillator. Thus, the oscillation frequency is given by the dispersion relation (7.9)

$$\omega^2 = \frac{C\,a\,(Bo + a^2)}{S}. \tag{7.52}$$

By taking $dE\,[(a,\,\omega\,(a))]/da = 0$, the necessary condition for minimum yields the neutral curve, that is, the critical values for sustained transverse waves:

$$E_c^T \approx -7.93\,(CS)^{3/4}, \tag{7.53}$$

$$\omega_c^{T^2} = 6\sqrt{5}C/S, \tag{7.54}$$

and

$$a_c^T = 1\Big/\sqrt{5}, \tag{7.55}$$

which correspond, respectively, to (7.11), (7.12), and (7.13) in the heat transfer problem. Thus it clearly appears that heat and mass transfer play a similar role and hence do excite the same type of transverse capillary–gravity waves when the Marangoni effect operates in the system.

7.3.3 Transverse Waves Not Obeying the Laplace Law

The numerical exploration of the problem posed in the preceding subsection shows that there exists a mode of transverse oscillation that does not follow from the Laplace–Kelvin equation $f\,(Bo) = 0$, and hence it cannot be considered as a standard capillary–gravity wave. From Equation (7.42) follows that

$$\frac{\partial^2 \zeta}{\partial t^2} + \frac{Bo + a^2}{SC}a\zeta = -2a^2\frac{\partial z}{\partial t} + 2a\frac{\partial w}{\partial z} - \frac{Ea^2}{S}\,(c - \zeta). \tag{7.56}$$

Then we have

$$\frac{\partial w}{\partial z} = \frac{f\,(Bo)\,(a - m) - 1}{2m - 2a - \lambda/a}\lambda\frac{\partial \zeta}{\partial t}$$

$$= \frac{a}{\lambda}\left[\lambda + 2ma + 2a^2 + f\,(Bo)\,(m + 1)\,\lambda\right]\frac{\partial \zeta}{\partial t}, \tag{7.57}$$

$$c - \zeta = -\frac{1}{q\lambda}\frac{\lambda f\,(Bo)\left(a + \frac{q - mS}{S-1}\right) - \lambda + 2a\frac{q-m}{S-1}}{2m - 2a - \lambda/a}\frac{\partial \zeta}{\partial t}$$

$$= \frac{a}{q\lambda^3}\,(m + a)\left[\lambda f\,(Bo)\left(a + \frac{q - mS}{S - 1}\right) - \lambda + 2a\frac{q - m}{S - 1}\right]\frac{\partial \zeta}{\partial t}. \tag{7.58}$$

Using these relations, Equation (7.56) becomes

$$\frac{\partial^2 \zeta}{\partial t^2} + \frac{Bo + a^2}{SC} a\zeta = -4a^2 \left[1 + \frac{a\,(a+m)}{\lambda}\right] \frac{\partial \zeta}{\partial t}$$

$$- 2af^2\,(Bo)\,(m+a)\frac{\partial \zeta}{\partial t} - \frac{Ea^3}{Sq\lambda^3}(m+a)^2$$

$$\times \left[\lambda f\,(Bo)\left(a + \frac{q-mS}{S-1}\right) - \lambda + 2a\frac{q-m}{S-1}\right]\frac{\partial \zeta}{\partial t}, \qquad (7.59)$$

where the high-frequency approximation $\omega >> a^2$ has been redundantly used.
This permits a simplified exploration of the problem and is still reasonable as we
search for transverse oscillatory modes that due to viscous damping penetrate
little in the liquid layer. Then Equation (7.59) becomes

$$\frac{\partial^2 \zeta}{\partial t^2} + \frac{Bo + a^2}{SC} a\zeta - \sqrt{2}\omega^{3/2}a^2 f\,(Bo)\,\zeta = -4a^2 \frac{\partial \zeta}{\partial t} - \sqrt{2\omega}a^2 f\,(Bo)\frac{\partial z}{\partial t}$$

$$+ \frac{Ea^3}{Sq\lambda}\left[1 - f\,(Bo)\frac{q - aS\sqrt{\lambda}}{S-1}\right]\frac{\partial \zeta}{\partial t}. \qquad (7.60)$$

Needless to say, Equation (7.60) yields the harmonic equation for capillary–
gravity waves (Laplace–Kelvin waves) when $f\,(Bo = 0)$.

Now let us explore the case $f\,(Bo \neq 0)$ [14]. Consider the case of low Schmidt
numbers $S << a^2/\omega$. Then it follows that

$$\frac{d^2 \zeta}{dt^2} + \left[4a^2 - \left(\frac{Ea}{2} - a^2\sqrt{2\omega}\right)f\,(Bo)\right]\frac{d\zeta}{dt} + \left(\frac{Bo + a^2}{SC}a - \frac{Ea^2}{S}\right)\zeta = 0. \quad (7.61)$$

This equation once more describes a harmonic oscillator. As done earlier, I search
for the neutral state. When the damping factor, i.e., the coefficient of the first
derivative with respect to time vanishes, one has

$$\frac{Bo + a^2}{SC}a - \omega^2 = \frac{Ea^2}{S}, \qquad (7.62)$$

$$\left(\frac{Ea}{2} - \sqrt{2\omega}a^2\right)\left(\frac{1}{a}\frac{Bo + a^2}{SC\omega^2}\right) - 4a^2 = 0. \qquad (7.63)$$

It follows that

$$E^2 - 2a\sqrt{2\omega}E + 8S\omega^2 = 0. \qquad (7.64)$$

There are two roots (both of them are positive):

$$E = a\sqrt{2\omega}\left(1 \pm \sqrt{1 - 4S\omega/a^2}\right) \approx a\sqrt{2\omega}\left(1 \pm 1 \mp \frac{2S\omega}{a^2}\right). \qquad (7.65)$$

It appears that the branch $E \approx 2a\,(2\omega)^{1/2}$ has no minimum when the wavelength
varies in the capillary length range. Therefore, it has no physical meaning and
presumably is a spurious consequence of the above–used simplifications. I consider the other root. Then the Marangoni number for neutral disturbances is

$$E = \frac{2\sqrt{2}S\omega^{3/2}}{a}, \tag{7.66}$$

with dispersion relation

$$\frac{Bo + a^2}{SC}a - \omega^2 = \frac{Ea^2}{S}, \tag{7.67}$$

thus showing that this wave mode is predicted for positive values of the Marangoni number, hence for the analogue heat transfer problem when the heating is done from the liquid side as for the onset of Bénard cells.

Equations (7.66) and (7.67) determine the neutral state. By minimizing, one obtains the critical (elasticity) Marangoni number,

$$E_c = \frac{5a^2 - Bo}{2aC}, \tag{7.68}$$

which permits transverse waves to be sustained at the without being controlled by the Laplace–Kelvin law $f(Bo = 0)$. Generally, the threshold for this mode of oscillation is higher than the threshold for steady cellular Bénard convection except when the capillary and the Schmidt (or Prandtl) numbers are small enough, as in the rather exotic case (albeit providing a possible experimental test) of a liquid He-4 layer near the lambda line. Note that to avoid buoyancy masking its observation the experiment should be conducted under low effective (micro) gravity conditions.

7.3.4 Longitudinal Waves (Lucassen Waves)

To avoid some tedious and rather not so relevant complications, I first simplify the problem. As the deformability of the surface has a negligible influence on the longitudinal wave motion I can safely set the capillary number C to infinity. Also, as the surfactant accumulation affects mainly high-frequency oscillations, and the frequency of longitudinal waves is of relatively not so high value, the number H can be neglected. With these assumptions Equations (7.39)-(7.43) reduce to

$$w = 0, \tag{7.69}$$

$$\frac{Ea^2}{S}c + \frac{\partial^2 w}{\partial z^2} = 0, \tag{7.70}$$

and

$$SH_z\frac{\partial c}{\partial t} = -\frac{\partial c}{\partial z}. \tag{7.71}$$

The surfactant concentration on the surface is chosen as the relevant variable in the harmonic oscillator. Differentiating Equation (7.38) with respect to z and using Equation(7.71), yields

$$-SH_z\frac{\partial^2 c}{\partial t^2} = \frac{\partial w}{\partial z} + S^{-1}\nabla^2\frac{\partial c}{\partial z}. \tag{7.72}$$

Then from Equation (7.69) it follows that

$$B = -A,$$ (7.73)

thus showing that here the rotational part is of the same order as the potential one near the surface. Accordingly, both rotational and potential terms are needed in the longitudinal wave case. Using Equation (7.73) and the long-wavelength assumption $a^2 << \omega << 1$, one gets

$$\frac{\partial w}{\partial z} \approx \frac{1}{m}\left(1 - \frac{a}{\sqrt{\lambda}}\right)\frac{\partial^2 w}{\partial z^2},$$ (7.74)

which using Equation (7.70) yields

$$\frac{\partial w}{\partial z} \approx Ea^2 q\left(\frac{\partial c/\partial t}{S^{3/2}\omega^2} - \frac{ac}{Sqm\sqrt{\lambda}}\right).$$ (7.75)

I also have

$$S^{-1}\frac{\partial}{\partial z}\nabla^2 c \approx q\frac{\partial c}{\partial t}.$$ (7.76)

Substitution of Equations (7.75) and (7.76) into (7.72) yields

$$-\frac{SH_z}{q}\frac{\partial^2 c}{\partial t^2} - \frac{Ea^3}{Sqm\sqrt{\lambda}}c = \left(1 + \frac{Ea^2}{S^{3/2}\omega^2}\right)\frac{\partial c}{\partial t}.$$ (7.77)

Note that

$$-\frac{SH_z}{q}\frac{\partial^2 c}{\partial t^2} \approx \frac{H_z\sqrt{S}}{\sqrt{2\omega}}\left(\frac{\partial^2 c}{\partial t^2} + \omega\frac{\partial c}{\partial t}\right)$$ (7.78)

and

$$-\frac{Ea^3}{Sqm\sqrt{\lambda}}c \approx \frac{Ea^3}{S\omega\sqrt{2S\omega}}\left(c + \frac{1}{\omega}\frac{\partial c}{\partial t}\right).$$ (7.79)

Replacing (7.78) and (7.79) into Equation (7.77) and keeping only the leading terms, one obtains

$$\frac{1}{\sqrt{2S\omega}}\left(SH_z\frac{d^2 c}{dt^2} - \frac{Ea^3}{S\omega}c\right) = -\left(1 + \frac{Ea^2}{S^{3/2}\omega^2}\right)\frac{dc}{dt},$$ (7.80)

which is an equation with the input–output energy balance written here for convenience in the right-hand side. At variance with transverse waves, the leading part in Equation (7.80) corresponds to the terms with first-order derivative, as

$$\left|\frac{1}{\sqrt{2S\omega}}\left(SH_z + \frac{Ea^3}{S\omega}\right)\right| << \left|\omega\left(1 + \frac{Ea^2}{S^{3/2}\omega^2}\right)\right|.$$ (7.81)

The potential part of Equation (7.80), though small compared with the dissipative one, determines the oscillating feature of this convective mode. In the neutral case when energy dissipation and energy supply given by the Marangoni effect compensate each other,

$$E = -S^{3/2}\omega^2 \big/ a^2 \quad or \quad \omega^2 = |E|\, S^{-3/2} a^2 \tag{7.82}$$

together with

$$E = -S^2 H_z \omega^3 \big/ a^3. \tag{7.83}$$

Thus, Equations (7.66) and (7.82) describe the characteristics of the longitudinal oscillation at the air–liquid interface and, clearly, E must take negative values,

$$E = -S^{1/2} H_z^{-1}. \tag{7.84}$$

The dispersion relation is

$$\omega^2 = a^2 S^{-1} H_z^{-2},$$

hence describing dispersionless waves that are necessarily dissipative as equation (66b) shows. E is non-vanishing, otherwise the longitudinal wave disappears. Figure 7.2 illustrates the dispersionless waves found experimentally [15] (see also [9]).

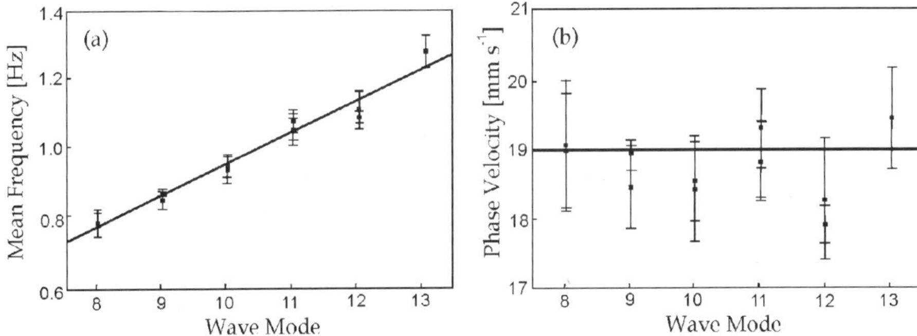

Fig. 7.2. Longitudinal waves excited by the Marangoni effect. Experimental data obtained when pentane vapor is adsorbed and, subsequently, absorbed by liquid toluene. (a) Dispersionless relation between frequency and wave number; (b) (phase) wave velocity. Error bars indicate maximum and minimum values observed [15].

7.4 Concluding Remarks

I have shown how a motionless horizontal liquid layer adsorbing a light surfactant (subsequently absorbed in the bulk) or, equivalently, being heated from the air side (and not from the liquid side as originally done by Bénard), hence stably stratified, can be made unstable with the Marangoni effect. Past a given threshold, the layer is bound to exhibit oscillatory motions (overstability) in the form of either transverse (capillary–gravity waves obeying or not the) or longitudinal motions in the form of dispersionless compression–expansion like waves. I have provided (linear) threshold values and the corresponding dispersion relations.

Comparison of Equations (7.8) or (7.51) and (7.82) shows that instability thresholds for both transverse and longitudinal motions depend on the Prandtl or the Schmidt number. This is to be expected for (inertial) oscillatory motions at variance with the onset of steady cellular Bénard convection which is insensitive to the values of these numbers (inertia has no significant role on this threshold).

Transverse waves are directly related to the interfacial deformation (measured by C) whereas longitudinal waves rather depend on Langmuir adsorption (measured by H_z). When these two waves are possible, cross-over from one to the other mode of instability may occur [6, 7, 8]. Indeed, in the cases discussed here depending on surface deformability (and surface tension) and/or surfactant adsorption, in a way such that when S is greater (respectively, smaller) than $C^3 / (7.93 H_z^2)^4$, longitudinal (respectively, transverse) waves are predicted to occur first.

A possibility not discussed here is the excitation of internal waves [8] (Brunt–Väisälä waves). Note that the actual liquid depth has played no role as I have used as a length scale the capillary length ($Bo = 1$). Thus I wonder how much of a role in experiments the layer depth plays. It is worth recalling that the shallow layer case tends to favor (relatively) long wavelength disturbances [4, 5].

Few clear-cut experimental results exist concerning the onset of wave motion occurring in the otherwise stably stratified motionless state. A truly systematic exploration of the phenomena is still pending. The available results [16, 17] obtained using n-Octane or n-Decane with heated air, and the pairs N_2/diphenyl or CO_2/diphenyl show threshold values of the Marangoni number on the order of 10^5 corresponding to thermal gradients somewhere in the range 10 to 10^2 K/cm. These critical values agree well with the predictions (7.11) or (7.53). Is there in experiment a smooth transition from the motionless state to (linear) overstability ending up in (weakly) nonlinear waves? Or is the transition abrupt? Noticeable is that experiments have been conducted when there is strong excitation of the system and, surely, the liquid layer was well above instability threshold, leading to nonlinear solitonic waves. A rich variety of results in such a case has been described in the past decade and full references are provided in [4]. Unfortunately, those experiments did not allow determination of threshold values as the phenomena appeared in a drastic and rather uncontrolled manner.

Acknowledgments

This research was supported by the Spanish Ministerio de Ciencia y Tecnologia under grant PB 96-599.

References

1. E.L. Koschmieder, *Bénard Cells and Taylor Vortices*, Cambridge University Press, Cambridge, UK, 1993.

2. P. Colinet, J.C. Legros, and M.G. Velarde, *Nonlinear Dynamics of Surface Tension-Driven Instabilities*, Wiley-VCH, New York, 2001.

3. M.F. Schatz and G.P. Neitzel, Experiments on thermocapillary instabilities, *Annu. Rev. Fluid Mech.* **33**, 93–127 (2001).

4. M.G. Velarde, A.A. Nepomnyashchy, and M. Hennenberg, Onset of oscillatory interfacial instability and wave motions in Bénard layers, *Adv. Appl. Mech.* **37**, 167–238 (2000) (and references therein).

5. A.A. Nepomnyashchy, M.G. Velarde, and P. Colinet, *Interfacial Phenomena and Convection*, Chapman & Hall-CRC, London, 2002 (and references therein).

6. A.Ye. Rednikov, P. Colinet, M.G. Velarde, and J.C. Legros, Two-layer Bénard–Marangoni instability and the limit of transverse and logitudinal waves, *Phys. Rev.* **E 57**, 2872-2884 (1998).

7. A.Ye. Rednikov, P. Colinet, M.G. Velarde, and J.C. Legros, Oscillatory instability and high–frequency wave modes in a Marangoni–Bénard layer with deformable freee surface, *J. Non-Equilib. Thermodyn.* **25**, 381–405 (2000).

8. A.Ye. Rednikov, P. Colinet, M.G. Velarde, and J.C. Legros, Rayleigh–Marangoni oscillatory instability in a horizontal liquid layer heated from above: coupling and mode-mixing of internal and surface dilational waves, *J. Fluid Mech.* **405**, 57–77 (2000).

9. J. Lucassen, Longitudinal capillary waves, Part 1, Theory, *Trans. Faraday Soc.* **64**, 2221–2229 (Part 1), and (Part 2), Experiments, *ibidem* **64**, 2230-2235 (1968).

10. P.L. García-Ybarra and M.G. Velarde, Oscillatory Marangoni-Bénard interfacial instability and capillary–gravity waves in single- and two-component liquid layers with or without Soret thermal diffusion, *Phys. Fluids* **30**, 1649–1655 (1987).

11. B.G. Levich, *Physicochemical Hydrodynamics*, Prentice-Hall, Englewood Cliffs, NJ (1965).

12. X.-L. Chu and M.G. Velarde, Sustained transverse and longitudinal waves at the open surface of a liquid, *Physicochem. Hydrodyn.* **10**, 727–737 (1988).

13. X.-L. Chu and M.G. Velarde, Sustained transverse and longitudinal waves induced and sustained by surfactant gradients at liquid–liquid interfaces, *J. Colloid Interface Sci.* **131**, 471–484 (1989).

14. X.-L. Chu and M.G. Velarde, Dissipative hydrodynamic oscillatoris. VI. Transverse interfacial oscillations not obeying the Laplace-Kelvin law, *Il Nuovo Cimento* **D 11**, 1631–1643 (1989).

15. A. Wierschem, H. Linde, and M.G. Velarde, Properties of surface wave trains excited by mass-transfer through a liquid surface, *Phys. Rev.* **E 64**, 022601 (2001).

16. L.Weh and H. Linde, Marangoni-stress-driven "solitonic" (periodic) wave trains rotating in annular container during heat transfer, *J. Colloid Interface Sci.* **187**, 159–165 (1997).

17. H. Linde, M.G. Velarde, A. Wierschem, W. Waldhelm, K. Loeschke, and A.Ye. Rednikov, Interfacial wave motions due to Marangoni instabilty. I. Traveling periodic wave trains in square and annular containers, *J. Colloid Interface Sci.* **188**, 16–26 (1997).

8 Hydrothermal Waves in a Disk of Fluid

Nicolas Garnier[†], Arnaud Chiffaudel[‡], and François Daviaud[‡]

†Laboratoire de Physique, CNRS UMR 5672, ENS-Lyon, 46 Allée d'Italie, 69364 Lyon CEDEX 07, France
‡DRECAM SPEC, CNRS URA 2464, CEA-Saclay, 91191 Gif-sur-Yvette CEDEX, France

Rayleigh–Bénard and Bénard–Marangoni instabilities have been studied for roughly a century and have served as prototypes for the transition to temporal chaos as well as spatio–temporal chaos of an initially . Using the Marangoni effect [1, 2] with a horizontal temperature gradient to drive the system out of equilibrium, one can observe propagating wave instabilities: hydrothermal waves [3]. This chapter presents different instability regimes of thermocapillary flows in extended geometry, focusing on propagating waves. We first introduce thermocapillary flows, and give some indications about physical effects involved. We then review experimental results in cylindrical geometry and illustrate how rich those systems are.

8.1 Thermocapillary Flows

The thermocapillary effect arises when a temperature gradient is applied to a fluid with a free surface [2, 4]. We consider here a disk of fluid with a free surface and a horizontal temperature gradient. The free surface is surrounded by ambient air.

8.1.1 Nondimensional Numbers

The fluid is characterized by its Prandtl number $Pr = \nu/\kappa$ which is the ratio of the diffusion coefficient of velocity to the one of temperature. But other numbers are important to describe the flow regime. We write ρ the density and σ the surface tension of the fluid. Those quantities depend on the temperature T, and one can define

$$\alpha = -\frac{1}{\rho}\frac{\partial \rho}{\partial T} \qquad \text{as well as} \qquad \gamma = -\frac{\partial \sigma}{\partial T}.$$

The thermal dilation coefficient α is always positive; γ is also positive for pure fluids: when the temperature increases, interactions between molecules decrease, and so does the surface tension.

The existence of a free surface, and therefore surface tension, implies that one can compute a capillary length λ_c. This quantity represents the spatial extent on which surface energy is comparable to bulk energy (e.g., gravity), that is it represents the spatial extent on which surface tension effects are relevant for a layer of fluid at rest. In the presence of gravity, the capillary length reads:

$$\lambda_{\mathrm{c}} = \sqrt{\frac{\sigma}{\rho g}},$$

where g is the magnitude of the gravity field.

When the fluid depth h is lower than λ_{c}, surface tension is predominant over gravity. Conversely, for h larger than λ_{c}, gravity is dominant. In fact, the ratio of h to λ_{c} is nothing less than the static Bond number Bo, also defined as the ratio of the surface tension forces to gravity:

$$Bo = \frac{\rho g h^2}{\sigma} = \left(\frac{h}{\lambda_{\mathrm{c}}}\right)^2$$

When the temperature is not uniform, buoyancy is present, represented by the Rayleigh number Ra as well as thermocapillarity, represented by the Marangoni number Ma. The Rayleigh number Ra is constructed as the ratio of buoyancy forces to viscous forces and Ma as the ratio of thermocapillary forces to viscous ones. If $\Delta T/l$ is the temperature gradient applied over distance l, they read

$$Ra = \frac{\alpha g h^4}{\nu}\frac{\Delta T}{\kappa l} \qquad \text{and} \qquad Ma = \frac{\gamma h^2}{\rho \nu \kappa}\frac{\Delta T}{l}.$$

The dynamical Bond number Bd is then defined as the ratio of thermocapillary forces to thermogravity forces:

$$Bd = \frac{Ra}{Ma} = \frac{\rho \alpha g h^2}{\gamma} = \left(\frac{h}{\lambda_{\mathrm{th}}}\right)^2$$

This defines another length scale $\lambda_{\mathrm{th}} = \lambda_{\mathrm{c}}\sqrt{\gamma/(\sigma\alpha)}$.

At ambient temperature (20°C), one has $\lambda_{\mathrm{c}} = 2.8\,\mathrm{mm}$ and $\lambda_{\mathrm{th}} = 88.5\,\mathrm{mm}$ for water. For the silicon oil we use, $\lambda_{\mathrm{c}} = 1.4\,\mathrm{mm}$ and $\lambda_{\mathrm{th}} = 3.0\,\mathrm{mm}$.

This dimensional analysis suggests the existence of different flow regimes. It turns out that those regimes are observed in the experiments as giving rise to different pattern–forming instabilities, as depicted in Figure 8.1.

For higher fluid depth $h > \lambda_{\mathrm{th}}$ (i.e., for thermogravity flows), one observes stationary pattern of rolls [5]; the axis of those rolls is parallel to the temperature gradient; those rolls have been observed in cylindrical geometry as well [6]. We are interested in wave patterns that appear in thermocapillary flows for $h < \lambda_{\mathrm{th}}$, among them are hydrothermal waves. Depending on the fluid depth, two types of hydrothermal waves are observed: type 1 (HW1) for medium fluid depth, and type 2 (HW2) for smaller fluid depth. Section 8.2 presents experimental observations of hydrothermal waves in the thermocapillary regime.

Finally, one also has to consider aspect ratios to discriminate if confinement is important. In the following, we assume large horizontal aspect ratios, that is, that the two horizontal directions (parallel and perpendicular to the temperature gradient) are much larger then the fluid depth h. In this case, confinement is negligible and we have an extended system in the two horizontal directions, as in the rectangular geometry of [7]. We are interested here in results obtained in

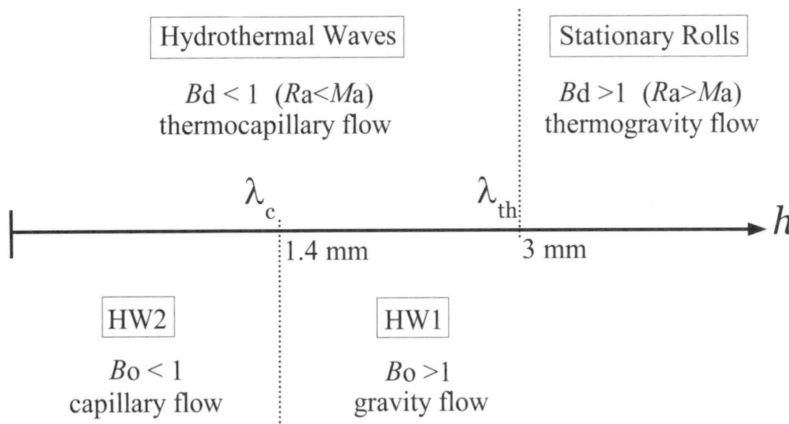

Fig. 8.1. Different regimes for thermocapillary and thermogravitational flow, depending on the fluid depth h. Limits are approximated by $h = \lambda_c$ and $h = \lambda_{th}$. We have identified instabilities as stationary rolls or hydrothermal waves that are observed experimentally in those different regimes. For hydrothermal waves, two types are observed (type 1, HW1, and type 2, HW2).

cylindrical geometry. The curvature is then an additional parameter that can be defined locally; as we discuss in Section 8.2.3, one of its effects is to localize the wave patterns [8].

8.1.2 Physical Mechanisms

Let's now briefly give an heuristic description of instability mechanisms in a fluid layer submitted to a vertical or horizontal temperature gradient. Oblique temperature gradients have effects close to horizontal ones [9].

Vertical Temperature Gradient: Bénard–Marangoni Instability

Pearson [10] gave a simple mechanism to explain hexagon formation in Bénard–Marangoni convection. In that case, the temperature gradient is purely vertical and the fluid is at rest when the system is on the thermodynamic branch. If one considers a positive temperature perturbation at the surface of the fluid, then one deduces that due to a locally smaller surface tension at that point, the fluid is flowing away from that point. Due to mass conservation, this implies that fluid is flowing up to the point at the surface, from the bulk which is at a higher temperature. So the perturbation is amplified: there is instability.

Horizontal Temperature Gradient

In that case, a basic flow exists when the system is on the thermodynamic branch, and we expect an instability into propagating waves. Giving a physical mechanism for propagating waves is a more tedious exercise than it is for stationary pattern. In the case of thermocapillary flows, the time-oscillatory nature of the instability comes as a result of the existence of well-defined profiles for the temperature and the velocity in the basic flow. A relevant mechanism has to involve features from those profiles such as the local sign of the horizontal and vertical temperature and velocity gradients. Smith [11] expressed two different mechanisms depending on the Prandtl number. Each of those is based on the Pearson mechanism, but whereas this former is Pr independent, Smith considered the extreme cases of a flow dominated by inertial effects ($Pr \to 0$) or by viscous effects ($Pr \to \infty$). The relaxation of temperature and velocity perturbations is then occurring on very different time scales. Depending on the signs of the underlying temperature and velocity gradients, an oscillatory behavior is shown to be unstable and to propagate along the horizontal temperature gradient (small Pr), or perpendicularly to it (large Pr).

It is worth mentioning that hydrothermal waves are an instability mode present in the absence of surface deflections, that is, assuming that the free surface is nondeformable. Taking into account surface deflections as in [12] may lead to another instability mode. Our experiments suggest that fluid depth variations are small compared to the fluid depth so that they can be neglected. The stability analysis of [3, 13] is then valid.

8.2 Experiments

We now present some experimental results, focusing on extended cylindrical geometries, with large horizontal aspect ratios. A sketch of the experimental cell is reproduced in Figure 8.2. The setup allows us to work at various fluid depths h while always having no meniscus on the side walls [14] and thus a perfectly homogeneous fluid depth h. We define the control parameter as $\Delta T = T_{\mathrm{ext}} - T_{\mathrm{int}}$. This quantity can be positive or negative, and both cases are not equivalent, due to the presence of curvature [8].

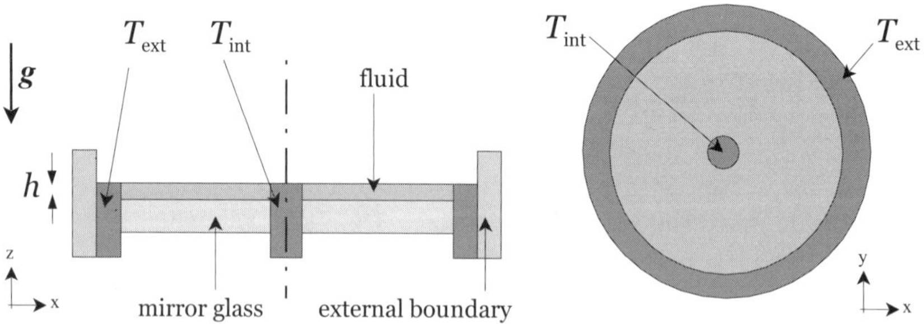

Fig. 8.2. Sketch of the experimental cell. External diameter is 135 mm, fluid depth h is of order 1 mm, and $\Delta T = T_{\mathrm{ext}} - T_{\mathrm{int}}$ is of order 10 K.

We use silicon oil of Prandtl number $Pr = 10$ and work with $h < \lambda_{\text{th}} = 3\,\text{mm}$ to have hydrothermal waves. Both cases $h \gtrless \lambda_{\text{c}} = 1.4\text{mm}$ are studied. Figure 8.3 gives a phase diagram of the experiment. Detailed observations and precise measurements have been performed for $h = 1.2\,\text{mm}$ (small Bo) and $h = 1.9\,\text{mm}$ (large Bo) for both positive and negative ΔT and are reproduced on Figure 8.3, left. The inset of Figure 8.3 (right) shows accurate determinations of HW1 and HW2 instability onsets for $\Delta T > 0$. The onset of each mode is determined by searching at which value of ΔT the squared amplitude of the corresponding pattern is vanishing, following a linear law. By doing so, we also check that each instability is supercritical.

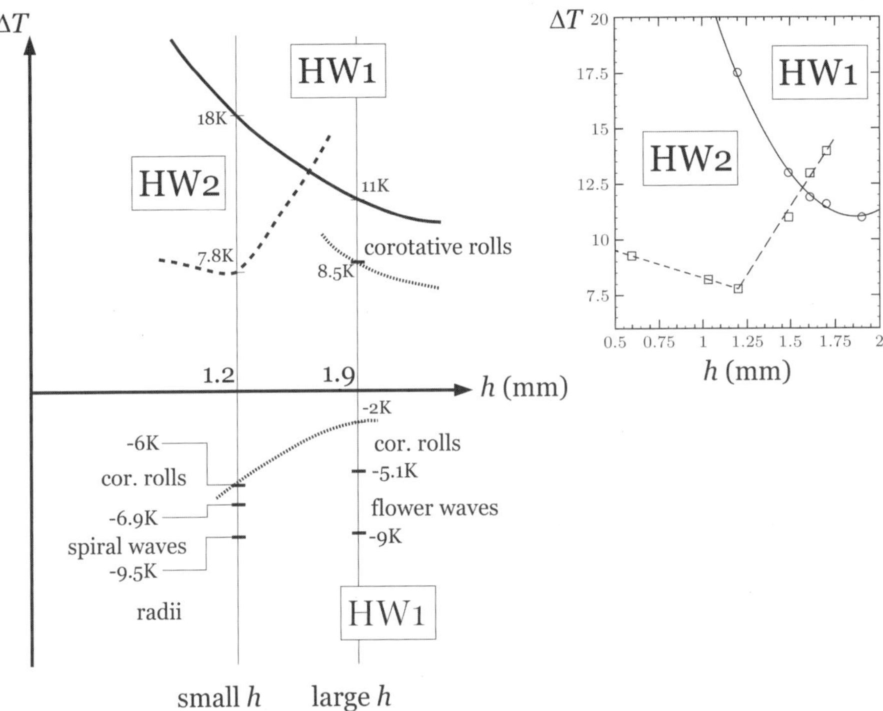

Fig. 8.3. Experimental phase diagram in a cylindrical cell. Left: General view. Right: Precise measurements of HW1 (o) and HW2 (□) onsets.

We report in Table 8.1 the experimental values of the critical temperature difference for the wave instabilities. The critical values of the Rayleigh Ra and Marangoni Ma numbers are also reported.

As can be seen in the phase diagram, the system exhibits a large variety of pattern-forming instabilities. This richness cannot be inferred from the dimensional analysis of the previous section. The following subsection details each of the observed structures.

Table 8.1. Critical Values at the Onset of Time-Oscillatory Patterns.

Fluid depth	$\Delta T_{\rm c}$	$Ra_{\rm c}$	$Ma_{\rm c}$	Pattern
$h = 1,2$ mm	7,8 K	80	500	HW2
	18 K	190	1150	HW1
$h = 1,9$ mm	11 K	735	1760	HW1
	$-5,2$ K	350	830	flowers
	-10 K	670	1600	HW1

8.2.1 $T_{\rm int} < T_{\rm ext}$

Large Bond Number (Large Fluid Depth)

For higher values of the Bond number, the basic thermocapillary flow composed of a single large roll is stable as long as $\Delta T < 8.5$ K. For $\Delta T > +8.5$ K a structuring of the basic flow occurs: concentric corotative rolls exist. These rolls first appear close to the hot side of the container, then invade the entire cell, as represented in Figure 8.4. For a higher temperature gradient ($\Delta T > +11$ K), hydrothermal waves appear. The corresponding pattern is composed of spiraling waves. Two realizations are presented in Figure 8.5. A source of waves with a large spatial extension may be present, as well as a (smaller) sink, and those objects separate two regions of right- and left-turning waves. On some realizations, a single wave (right- or left-turning) is present. In both cases, the two components of the local wavenumber are proportional; and their ratio is roughly constant anywhere in the cell and does not depend on the control parameter. Those hydrothermal waves are called HW1. They are also observed in rectangular geometries [5, 7, 15] and are well described by linear stability analysis [3, 13].

HW1 propagate with an angle from the temperature gradient. The radial propagation (i.e., the propagation along the temperature gradient) is always from the cold center ($T_{\rm int}$) towards the hot perimeter ($T_{\rm ext} = T_{\rm int} + \Delta T$). The orthoradial propagation is either to the right or to the left, both cases having equal probability.

Small Bond Number (Small Fluid Depth)

For lower values of the Bond number, and increasing the temperature difference ΔT from 0 K, no structuring of the basic flow by stationary corotative rolls is observed. The first instability mode appears for $\Delta T > 7.8$ K; it is a bidimensional hydrothermal wave (HW2) [14] localized near the center of the cell. At onset, the wavevector of the pattern is purely radial and the propagation is from the cold center towards the hot perimeter. As the control parameter is increased, the orthoradial component grows from zero. So the spatial structure of the HW2 mode evolves from a pulsing target to a spiraling pattern. Increasing the temperature gradient also results in an extension of the domain occupied by the HW2 pattern (Figure 8.6).

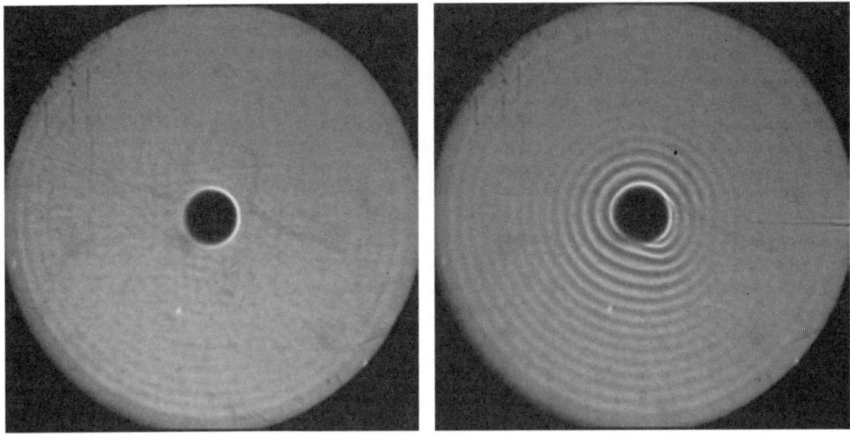

Fig. 8.4. Shadowgraph images for $h = 1.9\,\mathrm{mm}$ and cold center. Stationary corotative rolls appear on the hot side of the container (Left: $\Delta T = +10\,\mathrm{K}$), and invade the entire radial extension for higher temperature gradient (Right: $\Delta T = +11.75\,\mathrm{K}$).

Fig. 8.5. Shadowgraph images for $h = 1.9\,\mathrm{mm}$ and cold center. Left: $\Delta T = +13\,\mathrm{K}$, Right: $\Delta T = +14.25\,\mathrm{K}$. Hydrothermal waves of type I (HW1) in the shape of rotating spirals appear on top of the stationary rolls pattern. On the left, the pattern is composed of a uniform right-turning spiral.

For larger temperature gradients, HW1 appear in the whole domain of the cell where HW2 have a small or vanishing amplitude (Figure 8.7). Measurements of the local frequency and of the local wavenumber along the radial direction allow one to distinguish HW1 and HW2 instability modes. We have measured the onset of HW1 on top of the HW2 pattern, and shown that the two modes do not interact close to the onset of the second one (HW1). For a higher temperature gradient, interactions occur and the overall pattern is spatio–temporally chaotic.

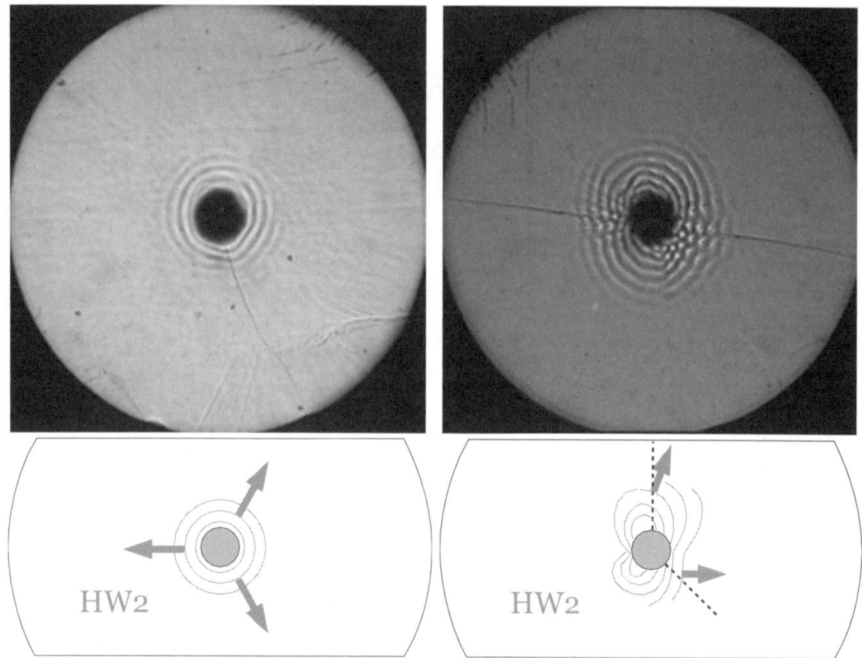

Fig. 8.6. $h = 1.2\,\text{mm}$ and cold center. Hydrothermal waves of type II (HW2). Close to onset (Left: $\Delta T = +8.5\,\text{K}$), the pattern is composed of pulsing targets and the wavevector is purely radial. Further in the supercritical region (Right: $\Delta T = +12\,\text{K}$), the wavevector has an additional azimuthal component. No stationary pattern is present. Arrows on the schematics represent the phase velocity.

8.2.2 $T_{\text{int}} > T_{\text{ext}}$

When the center is heated with respect to the outside perimeter, the dynamics is less coherent and more localized near the hot center; the phase diagram in the region $\Delta T < 0$ is richer.

Large Bond Number (Large Fluid Depth)

For $-2\,\text{K} < \Delta T < 0\,\text{K}$, the basic flow is stable. The first instability is stationary, and as in the case $\Delta T > 0$, it consists of a structuring of the base flow by corotative rolls. Those rolls appear on the hot side of the container, that is around the inner cylinder for $|\Delta T| > 2\,\text{K}$ (Figure 8.8, left).

For $|\Delta T| > 5.1K$, a time-oscillatory instability develops around the hot center. At onset, we observe a rotating hexagon (Figure 8.8, right). When the temperature gradient is increased, each corner of the hexagon moves away from the center, and the pattern then consists of a flower (Figure 8.9). While increasing $|\Delta T|$, one observes an elongation of the petals, then the apparition of a seventh petal (due to a modulational instability like the Eckhaus instability

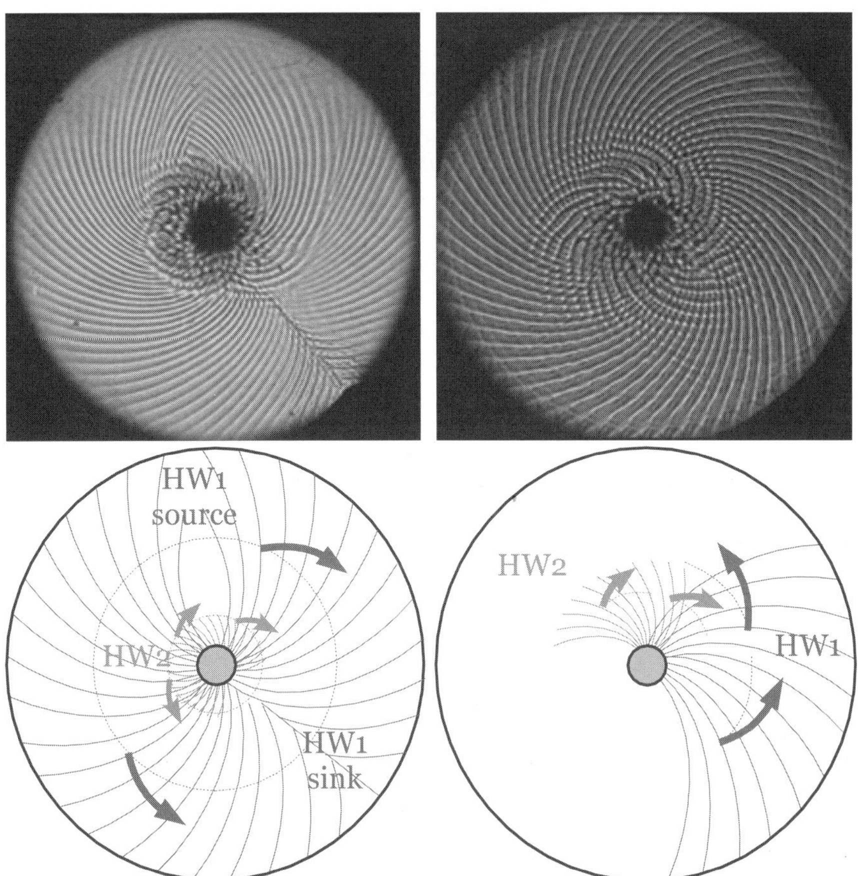

Fig. 8.7. $h = 1.2\,\mathrm{mm}$ and cold center, $\Delta T = +20\,\mathrm{K}$. HW2 are localized near the center. The HW1 pattern is either composed of right- and left-propagating waves (left photograph) or of a single wave (right photograph). Both configurations are unstable and the system oscillates randomly between the two regimes. The schematics detail propagation directions and source/sink positions for each realization. The radial propagation is always from the center and towards the perimeter.

in the azimuthal direction). Outside the flower, a structure is present with the same azimuthal wavenumber and it evolves with $|\Delta T|$ to form visible branches that rotate at the same angular frequency as the flower.

For $\Delta T < -9\,\mathrm{K}$, HW1 appear. Again, their radial propagation is from the cold side towards the hot side; so the HW1 pattern is such that energy flows from the external perimeter to the center. As we detail further, this situation is not comfortable and the structure has a strong tendency to be incoherent. Figure 8.10 presents a spatio–temporally chaotic realization for a large value of the control parameter. Stationary rolls are visible, and HW1 are barely recognizable.

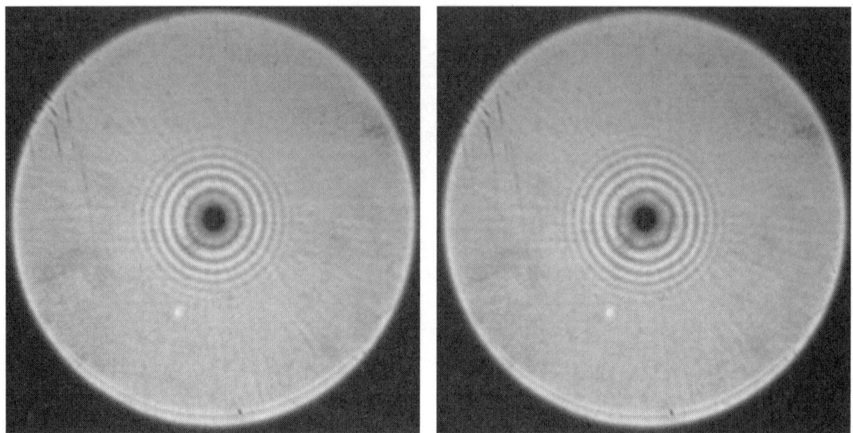

Fig. 8.8. $h = 1.9\,\mathrm{mm}$ and hot center. Left: $\Delta T = -5\,\mathrm{K}$, Stationary corotative rolls are present close to the hot center of the cell; Right: $\Delta T = -5.2\,\mathrm{K}$, same stationary rolls, with an additional hexagon turning around the center.

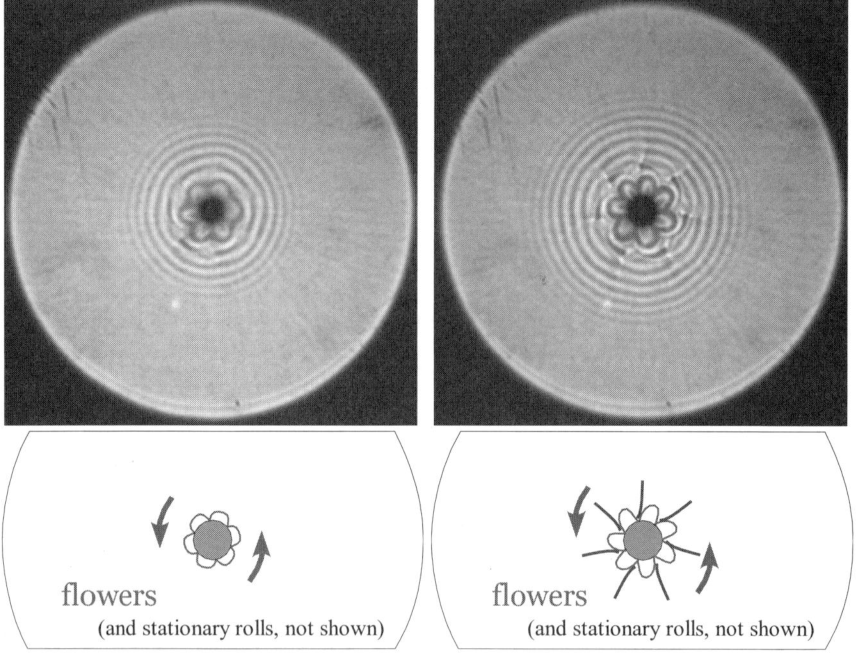

Fig. 8.9. $h = 1.9\,\mathrm{mm}$ and hot center; Left: $\Delta T = -5.6\,\mathrm{K}$, a six-petals flower is turning; Right: $\Delta T = -7\,\mathrm{K}$, an additional wavelength has appeared, as well as branches in between the petals, outside the flower.

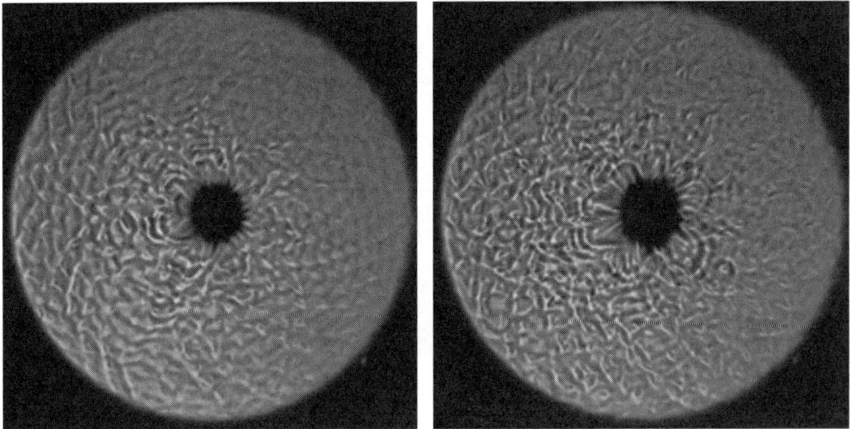

Fig. 8.10. $h = 1.9\,\text{mm}$ and hot center; Left: $\Delta T = -15.0\,\text{K}$; Right: $\Delta T = -20.0\,\text{K}$. Hydrothermal waves of type 1 have appeared on top of the flower pattern. Because the radial component of the HW1 wavevector is pointing towards the center of the cell, the spatial coherence of the resulting structure is small, and the overall pattern is spatio–temporally chaotic.

Small Bond Number (Small Fluid Depth)

First, for small $|\Delta T|$, only the basic flow is observed. For $\Delta T < -6\,\text{K}$, corotative rolls appear near the center. Their wavelength is small because they scale with the fluid depth which is small in that case. The amplitude of this stationary pattern is small and the corresponding shadowgraphic signal is very weak. Then, for $\Delta T < -6.9\,\text{K}$, spiral waves appear on top of the corotative rolls. We believe those are hydrothermal waves. Like HW2, they are localized near the center and their radial and azimuthal wavenumbers are not constant in space. An example is reproduced in Figure 8.11, left. The radial component of the wavevector is pointing towards the center, and the pattern is a left-turning wave. So the radial direction of propagation is reversed compared to an HW1 (or HW2) pattern of the same chirality obtained for $\Delta T > 0$.

For $\Delta T < -9.5\,\text{K}$, another propagating structure appears around the center of the cell. This structure has a wavevector that is purely azimuthal, so we label it "radii" as well. They are visible in Figure 8.11, right, very close to the inner cylinder. We observed that radii have a frequency close to twice that of hydrothermal waves. The azimuthal wavenumber is the same so the azimuthal phase velocity of the radii is twice that of the spirals. This allows us to conclude that they are different instabilities, although they may be 1:2 resonant. The strong localization of the radii suggests that it may be an instability of the hot boundary layer.

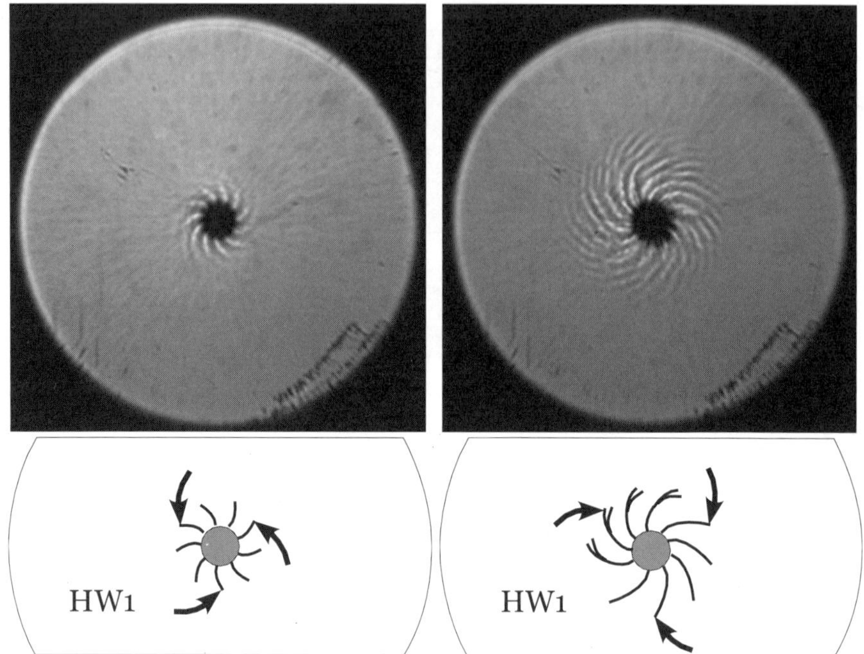

Fig. 8.11. $h = 1.2\,\mathrm{mm}$ and hot center. Left: $\Delta T = -8\,\mathrm{K}$, left-turning spiral waves; Right: $\Delta T = -11\,\mathrm{K}$, right-turning spiral waves and radii. Radial propagation of the spirals is from the external perimeter towards the inner plot. (Radii are not shown on the schematics.)

8.2.3 Curvature and Localization

Recent analytical and numerical work [17, 18] have also reported spiraling hydrothermal waves in cylindrical geometry, as well as other experimentally observed patterns. We can interpret to some extent the above observations using local curvature as introduced in [8]. The first effect of curvature is to distribute the temperature with a hyperbolic profile in the radial direction. This implies that the temperature gradient is larger in magnitude close to the center. For this reason, it is clear that the region close to the center becomes supercritical before the rest of the cell. This explains why most of the wave patterns we observe appear first close to the center, and afterwards in the bulk.

A second effect of curvature is to constrain the wavevectors. Close to the center, the azimuthal direction is not as extended as it is farther from the center. This implies that to keep a constant value of the wavenumber, the pattern has to increase the number of wavelengths in the azimuthal direction, and reduce the wavelength in the radial direction.

Moreover, in the case of hydrothermal waves, and in particular of HW1, the radial propagation is always from the cold to the hot side. In one–dimensional geometries [19, 20, 21], it has been checked that phase and group velocities point towards the same direction; this property seems satisfied as well in 2-D. When the center is colder than the outside, the pattern propagates from the center, and therefore spreads in the azimuthal direction. This can be achieved while keeping the spatial coherence. In the opposite case when the outside perimeter is colder than the center, the propagation is from the outside towards the inside, and the information (or energy) of the structure has to converge from an extended region to a confined one; in that case, any inhomogeneity of the structure (wavenumber, frequency, or amplitude) in the azimuthal direction will result in a destruction of the coherence of the converging process. For example, if the amplitude of the pattern is locally smaller at some given angle in the cell, the equivalent of a Bénard cell at that point will be squeezed by neighboring cells of larger amplitude, which will result in a decrease of the wavelength, and therefore of the amplitude: there is instability. This will ultimately result in the disappearance of the cell (i.e., of one wavelength), and the nucleation of modulations of, for example, the amplitude. Those modulations will be amplified by the same mechanism while the pattern converges towards the center. All those events occur incoherently in time and in space, giving rise to a turbulent pattern more easily when $T_{\mathrm{int}} > T_{\mathrm{ext}}$ than in the reverse case.

8.2.4 About Rectangular Geometries

We have also conducted experiments in rectangular geometry. In that case, positive and negative temperature gradients are equivalent, and one recovers the same behaviors: HW1 for high Bo and HW2 for low Bo. As in previous experiments [5, 15, 16], we found that corotative rolls appear only for large Bo, and prior to the HW1 instability; the HW1 are emitted by "line"-sources that extend over the whole extension between the hot and the cold sides. For smaller depth $h < \lambda_{\mathrm{c}}$, HW2 are observed; they are emitted by point sources located on the cold side of the container [5]. Moreover, for large fluid depth and therefore large Bond number, hydrothermal waves instability is replaced by the stationary instability into parallel rolls with the axis aligned with the temperature gradient, as presented in Figure 8.1.

8.3 Applications

Many quantitative results have been obtained for hydrothermal waves, on the theoretical side (linear stability analysis [3, 8, 9, 13]) as well as on the experimental side [4, 5, 7, 14, 15, 16]. For more fundamental studies, hydrothermal waves represent an ideal experimental nonlinear wave system. As for Bénard cells, hydrothermal waves are not just nice-looking; we have used them to study the transition to spatio–temporal chaos of a traveling wave system. In one horizontal dimension, they are well modeled by a complex Ginzburg–Landau equation [19, 20]. As their group velocity is finite, they are subject to the convective/absolute instability distinction [19, 21], not only for their primary onset, but also for the onset of their secondary instabilities. In a small supercritical range above onset, the hydrothermal wave pattern is advected out of the cell by the group velocity, in the very same way as a Rayleigh–Bénard stationary pattern is advected by an external flow.

One century after the pioneering work of Henri Bénard, thermocapillary flows are still a promising field of research not only from the hydrodynamical point of view, including challenging fundamental and applied industrial studies [1], but also as a robust model for the study of nonlinear waves and spatio–temporal chaos.

Acknowledgments

This work has been done in the *Groupe Instabilités et Turbulence*, SPEC, CEA Saclay. At the time of writing, N.G. was supported by a Joseph Ford fellowship from the *Center for Nonlinear Sciences*, Georgia Institute of Technology. Discussions with J. Burguete, H. Herrero, and G.P. Neitzel are acknowledged.

References

1. D. Kuhlman, *Thermocapillary Flows*, (Springer, New York, 1999).
2. S.H. Davis, Thermocapillary instabilities, *Ann. Rev. Fluid Mech.* **19**, 403 (1987).
3. M.K. Smith and S.H. Davis, Instabilities of dynamic thermocapillary liquid layers. Part 1. Convective instabilities, *J. Fluid Mech.* **132**, 119 (1983).
4. M.F. Schatz and G.P. Neitzel, Experiments on thermocapillary flows, *Annu. Rev. Fluid Mech.* **33**, 93 (2001).
5. J. Burguete, N. Mukolobwiez, F. Daviaud, N. Garnier, and A. Chiffaudel, Buoyant-thermocapillary instabilities in extended liquid layers subjected to a horizontal temperature gradient, Phys. Fluids **13**, 2773 (2001).
6. E. Favre, L. Blumenfeld, and F. Daviaud, Instabilities of a liquid layer locally heated on its free surface, *Phys. Fluids* **9**, 1473 (1997).
7. M.A. Pelacho *et al*, Local Marangoni number at the onset of hydrothermal waves, *Phys. Rev. E* **62**, 477 (2000).
8. N. Garnier and C. Normand, Effects of curvature on hydrothermal waves instability of radial thermocapillary flows, *C. R. Acad. Sci.*, Ser. IV **2** (8) 1227 (2001).
9. A.A. Nepomnyashchy *et al*, Stability of thermocapillary flows with inclined temperature gradient, *J. Fluid Mech.* **442**, 141 (2001).
10. J.R. Pearson, On convection cells induced by surface tension, *J. Fluid Mech.* **19**, 489 (1958).
11. M.K. Smith, Instability mechanisms in dynamic thermocapillary liquid layers, *Phys. Fluids* **29**, 3182 (1986).
12. M.K. Smith and S.H. Davis, Instabilities of dynamic thermocapillary liquid layers. Part 2. Surface-wave instabilities, *J. Fluid Mech.* **132**, 145 (1983).
13. J-F. Mercier and C. Normand, Buoyant-thermocapillary instabilities of differentially heated liquid layers, *Phys. Fluids* **8**, 1433 (1996).
14. N. Garnier and A. Chiffaudel, Two dimensional hydrothermal waves in an extended cylindrical vessel, *Eur. Phys. J.* **B19**, 87 (2001).

15. R.J. Riley and G.P. Neitzel, Instability of thermocapillary-buoyancy convection in shallow layers. Part 1. Characterization of steady and oscillatory instabilities, *J. Fluid Mech.* **359**, 143 (1998).
16. M.A. Pelacho and J. Burguete, Temperature oscillations of hydrothermal waves in thermocapillary-buoyancy convection, *Phys. Rev. E* **59**, 835 (1999).
17. S. Hoyas, H. Herrero and A. Mancho, Thermal convection in a cylindrical annulus heated laterally, J. Phys. A **35**, 4067 (2002).
18. S. Hoyas, H. Herrero and A. Mancho, Bifurcation diversity of dynamic thermocapillary liquid layers, *Phys. Rev. E* **66**, 057301 (2002).
19. N. Garnier, Ondes non-linéaires à une et deux dimensions dans une mince couche de fluide, Ph.D. Thesis, University Denis Diderot, Paris 7 (2000).
20. N. Garnier, A. Chiffaudel, and F. Daviaud, Nonlinear dynamics of waves and modulated waves in 1D thermocapillary flows. I. General presentation and periodic solutions, *Physica* **D174**, 1 (2003.)
21. N. Garnier, A. Chiffaudel, A. Prigent, and F. Daviaud, Nonlinear dynamics of waves and modulated waves in 1D thermocapillary flows. II. Convective/absolute transitions, *Physica* **D174**, 30 (2003).

9 Secondary Instabilities in Surface-Tension-Driven Bénard–Marangoni Convection

Kerstin Eckert and André Thess[*]

[1] Institute of Aerospace Engineering, Dresden University of Technology,
D-01062 Dresden, Germany, eckert@tfd.mw-tu-dresden.de
[2] * Department of Mechanical Engineering, Ilmenau University of Technology,
P.O. Box 100565, 98684 Ilmenau, Germany, thess@tu-ilmenau.de

We review recent experimental and theoretical findings on the behavior of surface tension dominated thermal convection in a thin fluid layer for Marangoni numbers significantly exceeding the threshold of the primary instability. Particular emphasis is placed on the description of a secondary instability which leads to a transformation of hexagonal convective cells into squares, referred to as the hexagon–square transition. Moreover, we explain the role of defects in the transition process and discuss some theoretical work aiming at the prediction of scaling laws for heat transport in the turbulent regime for low-Prandtl number fluids.

9.1 Introduction

When a horizontal layer of fluid with a free upper surface is heated from below, convective motion sets in as a result of two physical effects. The first one is the buoyancy force due to the temperature dependence of the density. The second one is the thermocapillary effect resulting from the temperature dependence of the surface tension, also called the Marangoni effect. When the fluid layer is sufficiently thin, thermocapillary effects dominate, and the influence of buoyancy can be neglected [1]. This case was studied by Bénard and is the central focus of the present work.

Since the first experiments of Henri Bénard [2, 3], surface-tension-driven convection in shallow liquid layers has been identified with the hexagonal planform, partially due to the suggestive power of aesthetic hexagonal patterns, and mostly due to the absence of reliable experimental and theoretical studies, extending sufficiently far into the nonlinear regime. This situation has been changed in the 1990s both due to focussed experimental efforts in optimized setups and to advanced computer resources allowing the direct numerical simulation of the governing equations. The present review intends to summarize these investigations, which have led to the discovery of a secondary instability, called hexagon–square transition, henceforth abbreviated to *HST*.

Surface-tension-driven Bénard convection is described by two parameters, the Marangoni number $M \equiv \gamma \Delta T d / \rho \nu \kappa$, which is the main control parameter, and by the Prandtl number $Pr = \nu/\kappa$, which is a property of the fluid. Here γ, ΔT, d, ρ, ν, and κ refer to the coefficient of surface tension variation, temperature difference across the fluid layer, thickness of the layer, density, kinematic viscosity, and thermal diffusivity of the fluid. It is known from linear stability analysis (Pearson [4] and Nield [5]) that the quiescent state of the fluid is stable as long as $M < M_c \approx 79.6$ and that convection sets in for $M > M_c$ in the form of hexagonal cells. The challenge of the past 15 years of research was to understand the behavior of Bénard convection far beyond this instability threshold.

How can one achieve a sufficiently high supercriticality ε defined as $\varepsilon = (M - M_c)/M_c$? One can employ either the thermocapillary or the solutocapillary effect. The latter rests on the concentration dependence of the surface tension rather than the temperature dependence.

Experimental studies of the interfacial convection during mass transfer of a surface-tension-lowering solute were performed indeed much earlier than comparative studies on the thermocapillary effect. We mention the contributions of Linde et al. [6, 7], Orell and Westwater [8] going back as far as the 1960s. The systematic investigations of Linde and coworkers [9, 10] have much advanced our understanding of surface-tension-driven pattern formation at high nonlinearity. In their picture of a hierarchic system of mostly nonstationary roll cells of different order, the stationary hexagonal Bénard cells appear as a very special case. They are observed when not too high supercriticality is accompanied by a stronger damping as imposed by the no-slip bottom condition in thin layers. Despite the advantage in achieving $\varepsilon \gg 1$, solutocapillary systems cannot compete with the thermocapillary analogue with respect to an exact study of the bifurcation of Bénard's hexagons. Because a steady solute feeding is difficult to realize, solutocapillary systems operate mostly under transient conditions which complicates the accurate determination of the Marangoni number. Furthermore, due to the small mass diffusivities a preferably linear stratification is difficult to generate.

Important contributions to a systematic study of the weakly nonlinear behavior of Bénard's hexagonal cells, such as the wavenumber evolution, originate from Koschmieder et al. [11, 12, 13, 14, 15]. With his very accurate experiments he laid the foundation for the later experiments discussed in Section 8.2.1. Among other things he introduced silicone oils as working fluids, due to their low sensitivity against interfacial contamination, or the sapphire window due to its superior heat conductivity, as top plate.

As following the recent developments on surface-tension-driven Bénard convection in detail goes beyond the scope of this chapter, we refer the reader to the excellent books or review articles by Davis [1], Koschmieder [16], Schatz and Neitzel [17], Colinet et al. [18], and Nepomnyashchy et al. [19].

9.2 Moderate Supercriticality

It seems appropriate to group our observations into two classes, corresponding to a moderately supercritical regime $0 < \varepsilon < 10$ and a highly supercritical regime

$\varepsilon > 10$. The present section, which is the main focus of the work, discusses the former case.

9.2.1 The Hexagon–Square–Transition

For $0 < \varepsilon < 4$ all experiments with a sufficiently large aspect-ratio (including Bénard's classical work) consistently lead to stable hexagonal pattern, which can be considered as the primary state of surface-tension-driven Bénard convection. A secondary instability of Bénard's hexagonal cells was discovered experimentally by Eckert (née Nitschke) and Thess ([20]; Eckert, Bestehorn, and Thess [21], the latter being henceforth abbreviated to EBT) and was later confirmed by Schatz, Van-Hook et al. (SVMSS) [22]. In parallel its existence has been verified in direct numerical simulations by Bestehorn [21, 23].

Both experimental groups have worked with controlled liquid–air-layer systems sandwiched between an isothermal bottom and a sapphire window on top whose temperatures (T_b and T_t) are precisely controlled to within ± 0.005 K or better. The ΔT used to quantify M and ε, respectively, is based on the conductive temperature drop ΔT_{cd} across the fluid layer assuming that both liquid and air layers are in conduction, independent of the actual value of $T_b - T_t$. Clearly, with onset of convection in the liquid, ΔT_{cd} refers to a fictional system. The advantage of ΔT_{cd}, however, is the possibility of external control by varying $T_b - T_t$, irrespective of the actual cellular pattern realized.

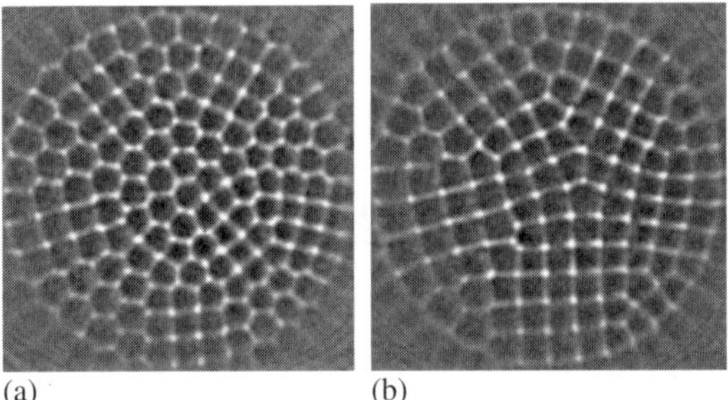

(a)   (b)

Fig. 9.1. (a) Example of a pattern of mixed symmetry at $\varepsilon = 4.65$ ($d = 1.67$ mm, $d_{air} = 0.34$mm, $Pr = 100$). (b) Typical square pattern at $\varepsilon = 5.3$ ($d = 1.56$ mm, $d_{air} = 0.46$ mm, $Pr = 100$) [24].

Silicone oils are used having moderate Prandtl numbers ranging between 81 (SVMSS) and 100 to 200 (EBT, [24]). To guarantee surface tension gradients as the main driving force, buoyancy effects must be kept small. The latter are characterized by the Rayleigh number $R \equiv \alpha g \Delta T d^3 / \nu\kappa$ with the liquid expansion coefficient α and the gravitational acceleration g. Thus, one requires the ratio M/R to exceed unity. This task is most easily realized in a shallow fluid layer. Eckert et al. [20, 21] used $d = (1.00 - 1.80)$ mm yielding $2 < M/R < 6$ and

SVMSS employed $d = (0.71 - 0.96)$ mm allowing for $8 < M/R < 15$. To exclude air convection, the thickness of the air layer must be small which in parallel minimizes the temperature difference lost across the air gap. Less than 0.47 mm and 0.10 mm were used in EBT and SVMSS, respectively. The aspect–ratio Γ defined as $\Gamma = L/d$, is moderate to large, and ranges between 20 and 65 where L denotes the diameter of the circular container.

The control parameter ε is slowly ramped during the experiments. EBT have chosen a time interval of about 4 h to achieve an ε-increase by $\Delta\varepsilon \sim 0.1$ and the experiments of SVMSS proceeded approximately eight times faster. The typical duration of the experiments is of the order of one to three weeks. For details of the experimental setup and procedure we refer the reader to the original papers.

Let us now focus on the question of what happens with the nearly ideal hexagonal patterns realizable in such experimental setups if supercriticality is increased. We begin with the well-founded results before discussing items that are not yet completely understood.

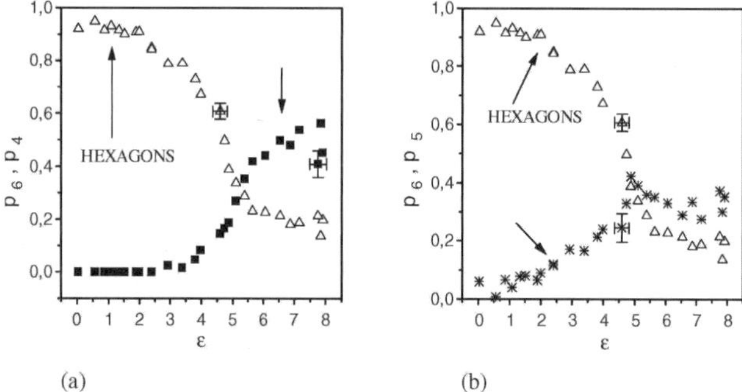

Fig. 9.2. Composition of the Bénard pattern: Fraction of hexagons, p_6, fraction of squares, p_4 (a), and fraction of pentagons, p_5 (b) as a function of ε ($d = 1.41$ mm, $d_{air} = 0.26$ mm, $\Gamma = 64$, $Pr = 100$) [21].

Experiments as well as numerical simulation provided clear evidence that hexagons lose their stability at a certain threshold ε_s whose value is discussed below. Beyond ε_s a state of mixed symmetry develops in which cells of hexagonal, pentagonal, and square planform coexist (Figure 9.1a). On increasing ε further, square cells begin to dominate until a state of nearly fourfold, square symmetry is achieved (Figure 9.1b).

The simplest way to quantify these compositional changes of the pattern as a function of ε is to introduce relative cell numbers $p_i = N_i/N$. They are defined as the ratio between the number of cells N_i of a given planform [hexagons (i=6), pentagons ($i = 5$), squares ($i = 4$)] to the total amount of complete cells N. The behavior of the p_i at $Pr = 100$ in a large-aspect-ratio-container ($\Gamma = 64$) is shown in Figure 9.2 for an experiment with increasing ε. At the onset of convection, the pattern comprises approximately 350 cells that are almost completely hexagonal, that is $p_6 = 0.98$. The deviation of p_6 from unity is due to a small amount of heptagonal and pentagonal cells that are unified in penta-hepta-defects.

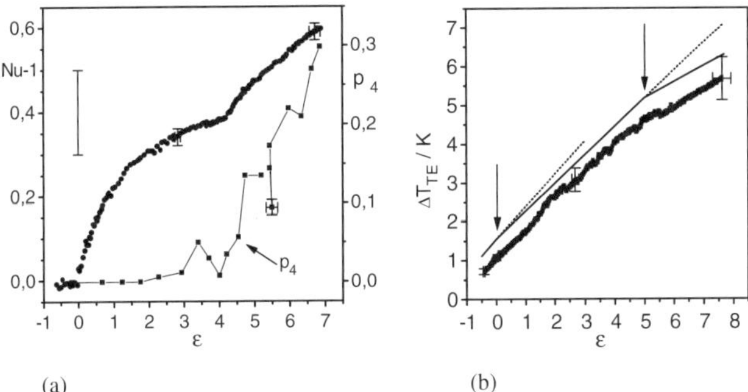

Fig. 9.3. (a) Nusselt number Nu and fraction of square cells p_4 as function of ε. ($d = 1.41$ mm, $d_{air} = 0.26$mm, $\Gamma = 56$, $Pr = 100$). (b) Temperature difference ΔT_{TE} in the liquid layer between bottom and a height $h = 0.87d$ ($d = 1.45$ mm, $d_{air} = 0.51$ mm, $\Gamma = 56$, $Pr = 100$). Convection sets in at $\Delta T_c = 1.08 K$ [21].

Up to $\varepsilon \sim 1$ the hexagon fraction p_6 can be kept as high as $p_6 = 0.96 \pm 0.03$ (Figure 9.2a). Beyond $\varepsilon \sim 2$, p_6 starts to decrease slowly at the expense of a growing pentagon number, p_5 (Figure 9.2b). These pentagons are no longer completely linked to heptagons but become more and more organized in the pentalines discussed later on. Beyond $\varepsilon = 4$, p_6 decreases drastically. In parallel, we detect a significant increase in p_4. The pentagon number p_5 grows, too, but at a slower rate, and reaches a maximum at $\varepsilon \sim 4.7$. Above this value, it again decreases slowly. At $\varepsilon \geq 6.5$ the p_i - values show minor changes only. An increase of p_4 above 55% was not observed in the aspect ratio under study ($\Gamma = 64$). With view to the larger p_i values of SVMSS this fact might be attributed to the lower value of M/R and to the higher Pr number. EBT and SVMSS performed a series of experiments in which ε is first increased until square cells dominate and then decreased again. They found that for decreasing ε a significant number of squares can be preserved below that ε_s at which the onset of square was detected for increasing ε. This hysteresis in the p_i-numbers reveals the subcritical nature of the transition. Experiments and numerical simulations in EBT provide a $\Delta\varepsilon = 0.5$ by which squares can be kept stable below ε_s.

To better understand the dynamics of the convective structure, measurements of the global heat flux and of local temperature differences, using thin thermo-couples, as function of ε have been performed by EBT. They detected bends in the slopes of both the injected electric power and the temperature difference between bottom and subsurface if the number of squares starts to rise. Trans-lated into nondimensional quantities, Nusselt number Nu and ε, these changes are plotted in Figure 9.3. With the onset of hexagonal Bénard cells the Nusselt

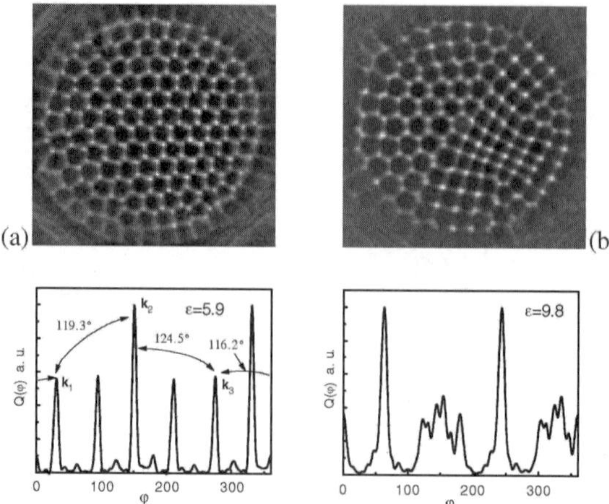

Fig. 9.4. Example for the onset of square cells via the propagation of a pentagon front at $Pr = 200$. (a) Distorted hexagonal pattern which is nearly free of other defects. Although the angle between modes $\mathbf{k}_1$ and $\mathbf{k}_2$ is equal to $\angle(\mathbf{k}_1, \mathbf{k}_2) \sim 120°$, $\angle(\mathbf{k}_2, \mathbf{k}_3)$ and $\angle(\mathbf{k}_3, \mathbf{k}_1)$ deviate from that value. This is visible in the orientational distribution obtained from radial integration of the power spectrum in Fourier space. (b) Square cells invade the hexagonal domain via propagation of a pentagon front ($d = 1.72$ mm, $d_{air} = 0.28$ mm; [24]).

number increases from unity. The second increase of the Nusselt number occurring in parallel with the rise of the p_4 number is remarkable. Any increase in the heat flux transferred by the cellular pattern slows down the increase of the temperature difference between bottom and surface in upstream regions, reflected by Figure 9.3b. Thus, both independently measured quantities show that with the appearance of square cells the heat flux through the layer increases. This capability of squares to transport more heat than hexagons was corroborated by the numerical simulations of Bestehorn on single hexagonal and square cells.

Another important and unexpected observation is the time-independence of both the square and mixed symmetry patterns, which has been proven by SVMSS. In experiments with a careful control of the lateral sidewall boundary condition they showed the stationarity of the patterns over timespans of several days. The stationary phase starts with the completion of a transient of less than 40 min after the ramping of ε is halted, in which the pattern activity relaxes. SVMSS demonstrated that any small inhomogeneity in the temperature or the surface pinning at the container wall causes a motion of adjacent cells parallel to the wall, inducing motions also in the interior of the layer. The stationarity of the pattern is in agreement with the simulations in EBT showing a time-dependent pattern for $Pr < 40$ only. It is in disagreement with the experiments of EBT where a slow spatio–temporal dynamics was observed with planform fluctuations $\Delta p_i = \pm 7\%$ over 20 h. In view of the finding of SVMSS we argue that indeed small pinning inhomogeneities along the rim of the container in EBT could be the reason for that difference.

Before looking into the details of the transition itself we first discuss ε_s and the wavenumber k, two quantities whose behavior still deserves a more detailed understanding.

EBT studied the wavenumber of hexagons and squares during runs with slowly increasing ε. During this type of evolution the wavenumber of the square cells exceeds that of the hexagons by about 8%. Interestingly, this behavior favors the squares by their higher perimeter-to-area-ratio which exceeds that of hexagons by more than 20% [25]. The wavenumber of both cell types decreases with increasing ε as expected from earlier works [14, 15, 26].

However, a different and rather peculiar behavior was found by SVMSS. They distinguished between ε-cycles in the reduced interval $0 < p_4 \leq 0.5$ and on the full interval $0 < p_4 \leq 1$. If SVMSS increased and decreased ε in the reduced interval, i.e., remaining fully in the mixed-symmetry-state, they found the average wavenumbers of hexagons and squares to be nearly identical and reproducible for each ε-cycle. The wavenumbers exhibit no hysteresis while hysteresis in the p_i-numbers is present. If, however, ε is cycled through the full interval, that is, comprising the hexagonal pattern and the nearly ideal square pattern, they found a significant hysteresis in the wavenumber associated with the formation of large squares which in turn induce large hexagons if ε is decreased again. Here, the occurence of p_i-number hysteresis depends on the initial conditions and on the prehistory of the pattern.

We are faced with analogous problems if we are to determine precisely the value ε_s for the onset of the transition. Both experiments deliver ε_s-values that are afflicted with a considerable scatter as can be anticipated from the foregoing discussion. EBT found $\varepsilon_s = 4.5 \pm 0.5$ based on linear fits of the p_4-function, and $\varepsilon_s = 4.2 \pm 0.3$ when the bend of the heat flux curves is used. SVMSS based their ε_s on reaching $p_4 = 0.5$ and provided $4.5 < \varepsilon_s^* < 6.4$. If we rescale this interval according to our definition of ε_s, based on the ε-value for which p_4 starts to rise (cf. EBT) we obtain $\varepsilon_s \approx 4.7 \pm 0.9$. Thus, both experiments provide comparable values. The large scatter in determining ε_s is not a problem of insufficient accuracy but appears to be an intrinsic feature of this transition. Both experiments showed that the onset of the *HST* depends on several factors such as the number of defects in the initial pattern, the history of the pattern, and the actual route by which square cells invade the pattern. Eckert [24] has distinguished between a quasi-isotropic route according to which domains of square cells are formed independently at five to six places within the container (cf. Figure 9.1a) and the less frequent front propagation route (Figure 9.4). The precondition for the latter route is the existence of a nearly perfect hexagonal pattern. On increasing ε this pattern can be preserved to rather high ε-values. The only defect that develops is a distortion of the pattern which shows in an increasing number of nonequilateral hexagons (Figure 9.4a). Beyond a certain ε_s, which can exceed that of the quasi-isotropic route by up to 30%, this distorted hexagonal pattern loses stability. A domain of square cells appears that is separated by a pentagon front from the hexagonal domain. The squares invade the latter one as the pentagon front propagates (Figure 9.4b). Although propagating fronts between hexagonal or roll pattern and the equilibrium state have been studied (cf. Nepomnyashchy et al. [19]), less is known about fronts separating squares and hexagons [27].

Until now we have left out the influence of the fluid properties on the *HST*. Indeed any increase in the Prandtl number enhances ε_s. The numerical simulations in EBT provide an estimation for the onset of the transition to occur at $\varepsilon_s = 0.28Pr^{0.68}$. For $Pr = 100$, the simulation predicts $\varepsilon_s^{sim} = 6.4 \pm 0.6$ which is 30% larger than the experimental value in EBT. For $Pr = 81$ a value $\varepsilon_s^{sim} = 5.6 \pm 0.6$ is obtained which lies in the transition range observed by SVMSS. The differences were discussed in EBT in terms of a reduced mean flow in the simulation, the neglect of buoyancy, and the different boundary conditions at the lateral walls. This fit demonstrates that a higher viscosity leads to a retardation of the *HST*. This was supported by Eckert's experiments [24] who found $\varepsilon_s = 6.5 \pm 1$ for $Pr = 200$ (Figure 9.4). This shift towards higher supercriticality is the reason why the extensive studies of Cerisier et al. [26, 28, 29, 30] at $O(Pr) \sim 1000$ revealed disordered hexagonal patterns only but no square cells. The classical defect of such a disordered hexagonal pattern is the Penta-Hepta-Defect (PHD, cf. Figure 9.5a). Fourier decomposition of the pattern (Ciliberto et al. [31]) has shown that the PHD is a bound state of two dislocations on two of the three roll subsystems, forming the hexagonal structure. The persistent time–dependence observed by Cerisier et al. [29] at high Pr in such disordered patterns is surprising in a twofold manner. On the one hand, the roll curvature, introduced by the dislocations bound to the PHD, of course gives rise to the production of vorticity. However, vertical vorticity ω_z scales with Pr^{-1}. According to EBT sufficiently low Pr-numbers are required to enable the action of ω_z as a kind of lubricant for the dislocation motion allowing the overcoming of the pinning effects of the small scale structure. On the other hand, the experiments of SVMSS, discussed above, brought evidence about the time-independence at a much lower Pr number where time-dependent effects are expected to be stronger. One possible source for the differences might be sought in the influences exerted by the buoyancy-driven convection in the noncontrolled air layer.

9.2.2 The Defect-Mediated Nature of the Transition

The ideas about the elementary structure of the defects as a specific ensemble of dislocations in the respective roll subsystems as well as about the action of ω_z, sketched above, are of help in describing details of the transition to square cells more accurately.

Both EBT and SVMSS discussed the transformation chain *hexagon → pentagon → square* occurring via shrinking of a cell edge to zero length implying the coalescence of two threefold vertices. Eckert and Thess [32] found that first this transformation can be traced back to a combined glide and climb motion of dislocations. Second, this type of motion is responsible for the systematic transformations in the defect shape that take place in the pattern on increasing ε

(cf. Figure 9.5). Nucleus of the entire transformation process is the penta-hepta-defect (Figure 9.5a). Above $\varepsilon \sim 2$, the heptagon of the PHD shows a tendency to diminish one cell side close to the pentagon until two of the seven cell knots coalesce. As a result, a so-called *(0,2)*-pentaline is formed, consisting of two pentagons and two nonequilateral hexagons (Figure 9.5b). *(0,2)* was used to describe that *non* pentagonal cells neighbor the *2*-pentagonal ones. The structure of the *(0,2)*-pentaline obtained from Fourier decomposition is shown in Figure 9.6. As in case of the PHD, the *(0,2)*-pentaline embedded in an otherwise hexagonal pattern is an ensemble of two dislocations in two of the three roll subsystems. The dislocation in roll set 1 with wave vector $\mathbf{k}_1$ (Figure 9.6a) was found close to the shortest side of the left non equilateral hexagon of the *(0,2)*-pentaline. The dislocation of roll set $\mathbf{k}_2$ is localized close to the central vertex[1] of the newly formed *(0,2)*-pentaline.

Remarkably, there is a separation distance between both dislocations of about two average hexagon side lengths.

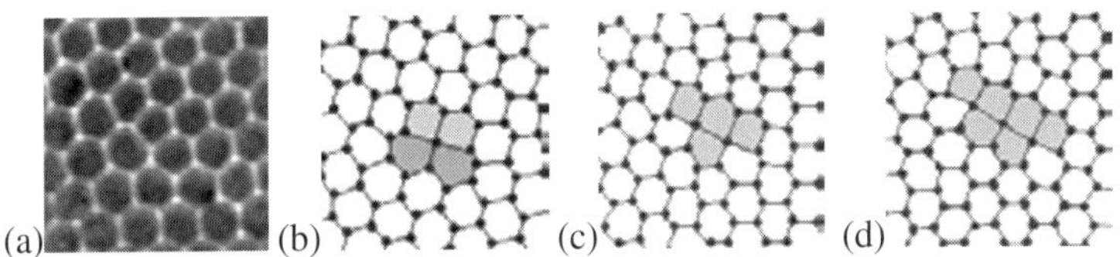

Fig. 9.5. Hierarchic system of defect transformation: (a) penta-hepta-defect; the pentagon is located near the center of the image; (b) *(0,2)*-pentaline; (c) *(1,3)*-pentaline; and (d) *(2,4)*-pentaline. (b) to (d) were obtained from the binarization of shadowgraphs. Pentagons are shown in light gray; after [32].

The existence of two *nonbound* dislocations is the basic difference between the *(0,2)*-pentaline and the PHD. Such a state is impossible in the weakly supercritical range because the dislocations become attracted towards each other (Rabinovich and Tsimring [33]). But in the moderately supercritical range, the *(0,2)*-pentaline is the dominant defect type which can be recognized already in the original photographs of Bénard's experiment (Figure 9.7). The formation of the *(0,2)*-pentaline and their subsequent evolution can be identified as the reasons for the rise of the pentagon number in Figure 9.2b.

What happens finally with a *(0,2)*-pentaline when ε is increased? We observe that this defect expands in the same manner as it was created. The expansion proceeds in analogy to the closing of a zip fastener. The shortest side of one of the two hexagons begins to shrink until both cell knots coalesce again. Consequently, two *new* pentagons are formed. The first one below the two pentagons of the *(0,2)*-pentaline and the other one beside them. In parallel, the hexagon adjacent to both new pentagons becomes non equilateral again. This ensemble, a

---

[1] This vertex has a coordination number four (four edges incident) as opposed to the other cell knots which are vertices of coordination number three.

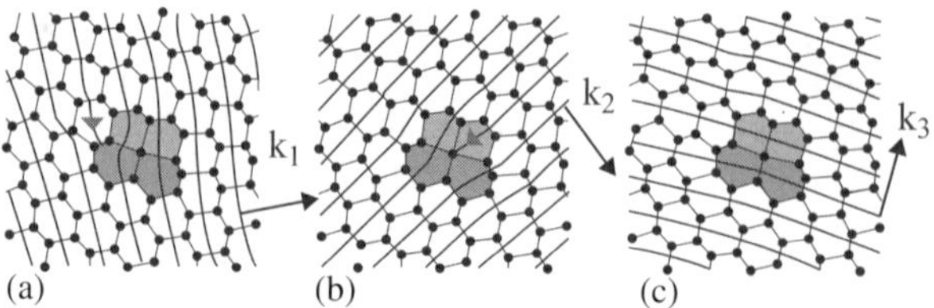

Fig. 9.6. Structure of a *(0,2)*-pentaline. The three roll subsystems with wavenumber $\mathbf{k}_i (i = 1, 2, 3)$ are superposed to the binarized shadowgraphs; after [32].

(1,3)-pentaline, is shown in Figure 9.5(c). The formation of two new pentagons is equivalent to a combined glide and climb motion of one dislocation, which includes a gliding by $2\pi/k_i$ together with a climbing by one to two hexagon side lengths. This mechanism is shown by means of the transformation of a *(0,2)-* into a *(1,3)*-pentaline in Figure 9.8. The mechanism of the pentagon formation consists of a combination of glide (steps *1* and *2*) and climb (step *3*) motions. The glide motion is initiated by the pinching off of a roll somewhere within the shown rectangle (step *1*). It is completed by the reformation of a new roll by the movement of the original dislocation (denoted by *A*) and the lower part of the pinched off roll towards each other (step *2*). The direction of the movement is indicated by arrows. The climb motion (step *3*) of the newly formed dislocation (denoted by *B*) finally causes the corresponding hexagon side to vanish. The next cycle of glide and climb motion is started by another pinching off denoted by *4*. The figure shows an enlarged detail of Figure 9.6(b).

Inspecting the surroundings of the rolls (Figures 9.6(a-b) or Figure 7 in [32]) we have noted that rolls going through pentagons are more strongly curved than rolls going through hexagons. A corresponding asymmetry in the mean flow can be expected which might explain the existence of the preferred direction of the combined glide and climb motions. This process continues via the appearance of a *(2,4)*-pentaline [Figure 9.5(d)], until finally the pentalines comprise 8 to 16 pentagons. In this advanced stage the formation of square cells between both pentagon cords of the pentaline is initiated (cf., e.g., Figure 9.1a). The process underlying the transformation of pentagons into squares is again the coalescence of neighboring vertices which proceeds in analogy to the transformation of hexagons into pentagons.

9.3 High Supercriticality and Outlook

A similarly fascinating issue as the HST itself is the question about the fate of the square Bénard cells at sufficiently high supercriticality. Preliminary evidence comes from Schatz [17] who showed a coarsening of the cells together with the formation of a substructure inside the cells. Considerably more material is available from mass-transfer systems where much higher supercriticality can be realized related with the low solute diffusivities. As mentioned in Section 8.1

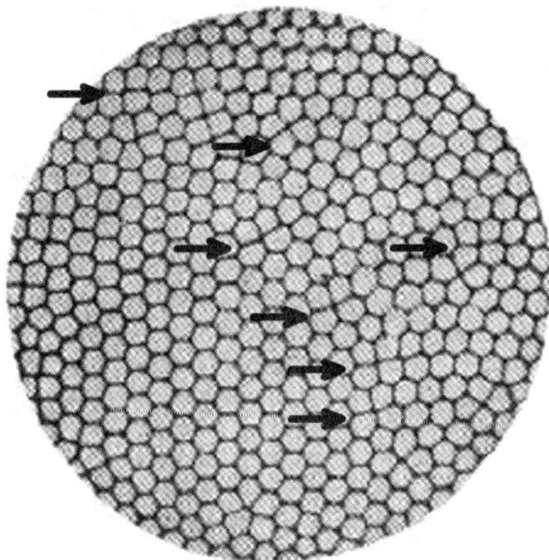

Fig. 9.7. A reproduction of one of Bénard's original photographs [3] displaying numerous *(0,2)*-pentalines marked by arrows.

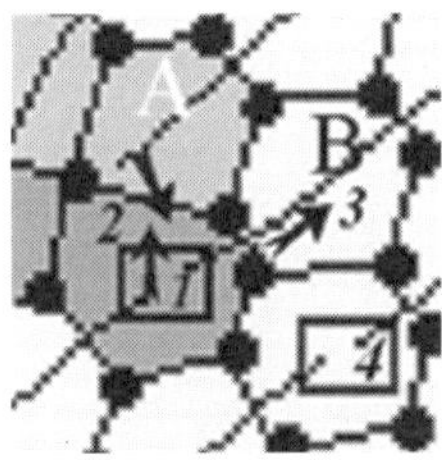

Fig. 9.8. The mechanism of the pentagon formation: Glide motion (steps *1* and *2*) of the original dislocation denoted by *A* and climb motion (step *3*) of the new dislocation *B*. The direction of the movement is indicated by arrows [32].

Linde et al. [9, 10] observed a hierarchic system of roll cells of first-, second- and higher-order where the roll cells, e.g., of second order, are built up by the one-order-in-magnitude smaller roll cells of first order traveling along the interface. With increasing supercriticality the wavenumber spectrum becomes broader and the cellular structure more and more disordered. Although experimental studies of the thermocapillary regime with $\varepsilon > 10$ in high-Prandtl number fluids are still lacking, numerical simulation (Thess and Orszag [34, 35]) indicates that strong temperature gradients develop for sufficiently high Marangoni number between the convective cells. This fact is illustrated in Figure 9.9, which shows the surface temperature of infinite-Prandtl-number convection for $M = 2000$.

In addition to the behavior of high-Prandtl-number fluids (silicone oils), liquid metals, characterized by a very low Prandtl number, have attracted the attention of researchers. On the one hand, this is due to the fundamental interest in low-Prandtl-number convection. On the other hand, material processing techniques such as laser welding and electron beam evaporation involve surface-tension-driven flows at Marangoni numbers sometimes as high as $M = 10^5$ and require a thorough understanding of the fully developed turbulent regime. High-resolution direct numerical simulations (Boeck and Thess [36, 37, 38]) have helped to gain insight into this region of parameter space and to check the validity of phenomenological scaling models for the Nusselt number as a function of the Marangoni number proposed by Pumir and Blumenfeld [39] and Karcher et al. [40].

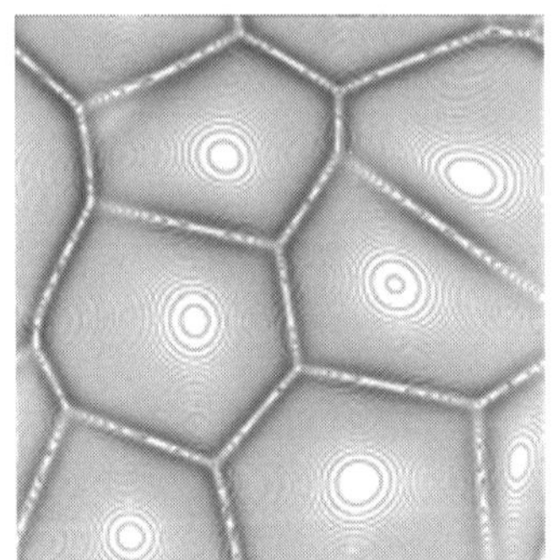

Fig. 9.9. Horizontal Laplacian applied to the surface temperature field. Marangoni number is $M = 2000$, aspect ratio $\Gamma = 3$, spatial resolution 128^2 x 64 [35].

We should mention finally that large portions in the parameter space of surface-tension-driven Bénard convection, such as the influence of the Biot number or the ratio M/R of Marangoni to Rayleigh number, still remain unexplored. Also, the exploration of the band of stable wavenumbers has started until recently by application of a thermal laser writing technique [41]. Despite the progress in weakly nonlinear theory and direct simulation, the problem of why hexagons or squares are the preferred planforms is not yet understood in a physically transparent way. We hope that this text will stimulate further research on this fundamental problem in fluid dynamics. So, the analysis of similarities to square cells found in other systems, such as ferrofluids [42], Rayleigh–Bénard convection (RBC) in binary mixtures [43] or RBC affected by temperature-dependent viscosity [44] or rotation (cf. Chapter 4 by G. Ahlers in this volume) might deliver deeper insights. Furthermore, new generations of experiments, for example, using substances near their critical point or high-purity liquid metals could help to improve our understanding of Bénard convection in the coming century as much as silicone oils did in the previous century after Bénard's seminal work.

References

1. S.H. Davis, Thermocapillary instabilities, *Annu. Rev. Fluid Mech.* **19**, 403–435 (1987).
2. H. Bénard, Les tourbillons cellulaires dans une nappe liquide, *Rev. Gén. Sci. Pures Appl.* **11**, 1261–1271 (1900).
3. H. Bénard, Les tourbillons cellulaires dans une nappe liquide transportant de la chaleur par convection en régime permanent, *Annales Chimie et de Physique* **23**, 62–144 (1901).
4. J.R.A. Pearson, On convection cells induced by surface tension, *J. Fluid Mech.* **4**, 489–500 (1958).
5. D.A. Nield, Surface tension and buoyancy effects in cellular convection, *J. Fluid Mech.* **19**, 341–352 (1964).
6. H. Linde, Untersuchungen über den Phasenübergang an ebenen und waagerechten flüssig-flüssig Phasengrenzen bei verschiedenartigen Einsatz grenzflächenaktiver Stoffe (In German), *Fette, Seifen und Anstrichmittel* **6**, 1053–1056 (1958).
7. H. Linde and K. Loeschke, Rollzellen und Oszillationen beim Wärmeübergang zwischen Gas und Flüssigkeit, *Chem. Ing. Tech.* **39**, 65–74 (1966).
8. A. Orell and J.W. Westwater, Spontaneous interfacial cellular convection accompanying mass transfer: ethylene glycol-acetic acid-ethyl acetate, *A.I.Ch.E. J.* **8**, 350–356 (1962).
9. H. Linde, P. Schwartz and H. Wilke *Dissipative Structures and Nonlinear Kinetics of the Marangoni Instability*, in *Lecture Notes in Physics* **105** (T.S. Sorensen, Ed.), Springer, Berlin (1979).
10. H. Linde in *Convective Transport and Instability Phenomena*, J. Zierep and H. Oertel (Eds.), Braun, Karlsruhe (1982).
11. E.L. Koschmieder, On convection under an air surface, *J. Fluid Mech.* **30**, 9–15 (1967).
12. E.L. Koschmieder and M.I. Biggerstaff, Onset of surface-tension-driven Bénard convection. *J. Fluid Mech.* **167**, 49–64 (1986).
13. E.L. Koschmieder and S.A. Prahl, Surface-tension-driven Bénard convection in small containers, *J. Fluid Mech.* **215**, 571–583 (1990).
14. E.L. Koschmieder, The wavelength of supercritical surface-tension-driven Bénard convection, *Eur. J. Mech. B/Fluids* **10**, 233–237 (1991).
15. E.L. Koschmieder and D.W. Switzer, The wavenumber of supercritical surface-tension-driven Bénard convection, *J. Fluid Mech.* **240**, 533–548 (1992).
16. E.L. Koschmieder, *Bénard cells and Taylor vortices,* Cambridge University Press, UK (1993).
17. M.F. Schatz and G.P. Neitzel, Experiments on Thermocapillary Instabilities, *Annu. Rev. Fluid Mech.* **33**, 93–129 (2001).
18. P. Colinet, J.C. Legros and M.G. Velarde, *Nonlinear Dynamics of Surface-Tension-Driven Bénard Convection,* Wiley-VCH, New York (2001).
19. A.A. Nepomnyashchy, M.G. Velarde and P. Colinet, *Interfacial Phenomena and Convection.* Chapman and Hall/CRC, Boca Raton, Fl (2001).
20. K. Nitschke and A. Thess, Secondary instability in surface-tension-driven Bénard convection. *Phys. Rev. E* **52**, R5772–R5775 (1995).
21. K. Eckert, M. Bestehorn and A. Thess, Square cells in surface-tension-driven Bénard convection: experiment and theory. *J. Fluid Mech.* **356**, 155–97 (1998).
22. M.F. Schatz, S.J. Van-Hook, W.D. McCormick, J.B. Swift and H. Swinney, Time-independent square pattern in surface-tension-driven Bénard convection, *Phys. Fluids* **11**, 2577–2582 (1999).

23. M. Bestehorn, Square patterns in Bénard-Marangoni convection, *Phys. Rev. Lett.* **76**, 46–49 (1996).
24. K. Eckert, From hexagonal towards square convection cells: experimental studies of surface-tension-driven Bénard convection (In German), Ph.D. Thesis, Otto-von-Guericke University, Magdeburg, Germany (1998).
25. U. Thiele and K. Eckert, Stochastic geometry of polygonal networks: an alternative approach to the hexagon-square transition in Bénard convection, *Phys. Rev. E* **58**, 3458–3468 (1998).
26. P. Cerisier, C. Pérez-García, C.,Jamond and J. Pantaloni, Wavelength selection in Bénard-Marangoni convection, *Phys. Rev. A* **35**, 1949–1952 (1987).
27. C. Kubstrup, H. Herrero and C. Pérez-García, Fronts between hexagons and squares in a generalized Swift-Hohenberg equation, *Phys. Rev. E* **54**, 1560 (1996).
28. P. Cerisier, C. Jamond, J. Pantaloni and C. Pérez-García, Stability of roll and hexagonal patterns in Bénard-Marangoni convection, *Phys. Fluids* **30**, 954–959 (1987).
29. P. Cerisier, R. Occelli, C. Pérez-García and C. Jamond, Structural disorder in Bénard-Marangoni convection. *J. Phys.* (France) **48**, 569–576 (1987).
30. P. Cerisier, C. Pérez-García and R. Occelli, Evolution of induced patterns in surface tension-driven Bénard convection, *Phys. Rev. E* **47**, 3316–3325 (1993).
31. S. Ciliberto, E. Pampaloni and C. Pérez-García, Competition between different symmetries in convective pattern, *Phys. Rev. Lett.* **61**, 1198–1201 (1988).
32. K. Eckert and A. Thess, Nonbound dislocations in hexagonal patterns: Pentagon lines in surface-tension-driven Bénard convection, *Phys. Rev. E* **60**, 4117-4124 (1999).
33. M.L. Rabinovich and L.S. Tsimring, Dynamics of dislocations in hexagonal patterns, Phys. Rev. E **49**, R35–R38 (1994).
34. A. Thess and S.A. Orszag, Temperature spectrum in surface-tension-driven Bénard convection, *Phys. Rev. Lett.* **73**, 541–544 (1994).
35. A. Thess and S.A. Orszag, Surface-tension-driven Bénard convection at infinite Prandtl number, *J. Fluid Mech.* **283**, 201–230 (1995).
36. T. Boeck and A. Thess, Inertial Bénard-Marangoni convection, *J. Fluid Mech.* **350**, 149–175 (1997).
37. T. Boeck and A. Thess, Turbulent Bénard-Marangoni convection: Results of two-dimensional simulations. *Phys. Rev. Lett.* **80**, 1216–1219 (1998).
38. T. Boeck and A. Thess, Bénard-Marangoni convection at low Prandtl number. *J. Fluid Mech.* **399**, 251–275 (1999).
39. A. Pumir and L. Blumenfeld, Heat transport in a liquid layer locally heated on its free surface, *Phys. Rev. E* **54**, 4528–4531 (1996).
40. C. Karcher, R. Schaller, T. Boeck, C. Metzner and A. Thess, Turbulent heat transfer in liquid iron during electron beam evaporation, *Int J. Heat Mass Transfer* **43**, 1759–1766 (2000).
41. D. Semwogerere and M.F. Schatz, Evolution of hexagonal patterns from controlled initial conditions in a Bénard-Marangoni convection experiment, *Phys. Rev. Lett.* **88**, 054501/1–4 (2002).
42. B. Abou, J.E. Wesfreid and S. Roux, The normal field instability in ferrofluids: hexagon-square transition mechanism and wavenumber selection, *J. Fluid Mech.* **416**, 217 (2000).
43. P. Le Gal, A. Pocheau and V. Croquette, Square versus roll pattern at convective threshold, *Phys. Rev. Lett.* **54**, 2501–2504 (1985).
44. D.B. White, The planforms and onset of convection with a temperature-dependent viscosity, *J. Fluid Mech.* **191**, 247–286 (1988).

Part IV

Bénard and Wakes

10 Wake Instabilities Behind Bluff Bodies

Michel Provansal

Institut de Recherche sur les Phénomènes Hors Equilibre
U.M.R. 6594 C.N.R.S. Technopôle de Château–Gombert
49 rue F. Joliot-Curie, B.P. 146 13384 Marseille cedex 13 France
michel.provansal@irphe.univ-mrs.fr

The observation by Bénard of a vortex street in the wake of a circular cylinder has been commonly associated with the stability analysis of the double alternate street proposed by von Kármán. After a short historical review of these studies, we present the main progress in understanding this instability during the last decade. New experiments and the control of two-dimensional flows have clarified the source of different modes as well as the results of three-dimensional numerical simulations. The introduction of new concepts (absolute and convective instabilities) allows us to link the velocity defect in the wake to the mechanism of formation of the vortex street. The dynamics of the wake has been successfully compared to the nonlinear Landau and Ginzburg–Landau models. Besides the configuration of the circular cylinder wake, we describe different kinds of bluff bodies and the diversity of their applications.

10.1 Introduction and Historical Aspects

Although the frequency of the sound emitted by a cylindrical rod translating in air has been studied since 1878, after Strouhal paper [1], the observation of the vortex street and the measurements of its geometrical characteristics were reported for the first time by Henri Bénard in two different papers, in November 1908 [2, 3] (a"qualitative" study was previously published by A. Mallock in 1907 [2]). He described, "Les tourbillons produits périodiquement, se détachent alternativement à droite et à gauche du remous d'arrière ...pour former une double rangée d'entonnoirs stationnaires, ceux de droite dextrogyres, ceux de gauche lévogyres, séparés par des intervalles égaux ("The periodic vortex are shed alternately on the right and on the left of the to form a double row of steady funnels, separated by the same distances"). Also see Fig. 10.1 where there is a wrong rotation of the vortices in the upper row. An ingenious setup, presented in the second note of 1908 [3], permitted to Bénard [5] to film the movement of the free surface by an optical method and cinematography. A careful investigation of the best views, measuring with a 5 μm resolution the location of each vortex on the free surface at each period, gave the spatio-temporal structure of the wake (cf. Figure 10.2a and 10.2b). The first paper of 1908 [2] is now well known but we should emphasize that this subject has attracted Bénard's attention from 1908 to 1927 in Lyon [6, 7, 8, 9] (see Chapter 1 by J.E. Wesfreid in this volume). His results are reported in at least seven different papers. Different experiments

Quand les vitesses de rotation sont faibles (faible vitesse de l'obstacle ou forte viscosité du liquide, et, dans tous les cas, tourbillons près de s'éteindre), les entonnoirs sont sensiblement de révolution. La méthode optique, qui équivaut à dessiner le relief de la surface liquide en l'éclairant sous une incidence presque rasante, les enregistre par des *croissants* demi-circulaires, estompés à l'intérieur (partie supérieure de la figure, à droite).

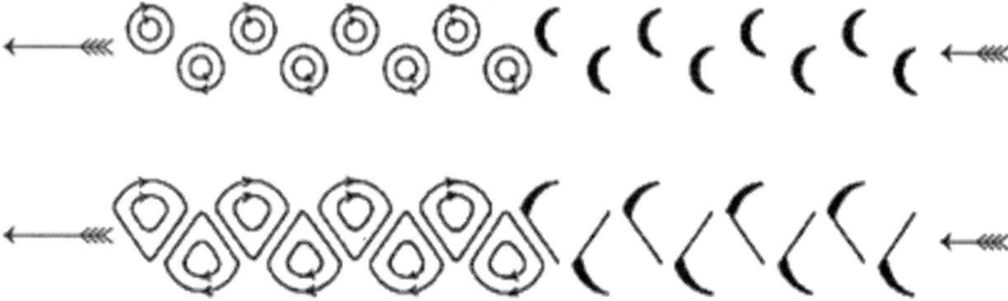

Quand les vitesses de rotation sont plus grandes (grande vitesse de l'obstacle ou faible viscosité du liquide), les entonnoirs des deux rangées se déforment mutuellement; leur contour prend une forme de raquette; les cavités sont plus abruptes du côté intérieur et leur bord y est rectiligne. Si la lumière quasi rasante arrive par exemple dans la direction indiquée par la flèche pointillée, ou à 180°, elle dessine les cavités en forme de *faux* (partie inférieure de la figure, à droite) (¹).

Fig. 10.1. Extract of Bénard's observation [2].

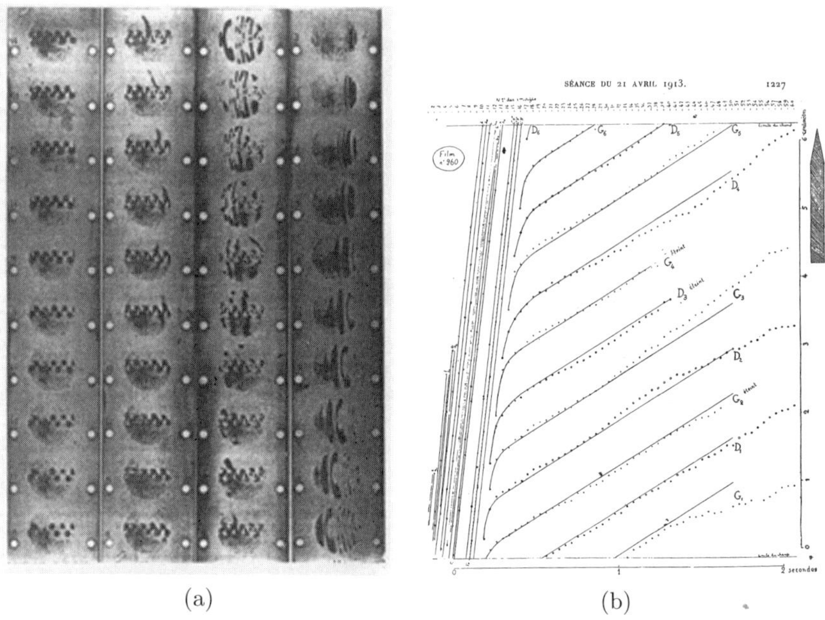

(a)   (b)

Fig. 10.2. (a)Pictures and (b) curves measured by H. Bénard in 1913 [5].

were performed by Riabouchinsky in water and in air [10] and von dem Borne [11] visualized vortices in air with tobacco smoke. Theodore von Kármán [12] wrote "that, at the same time, Prandtl had a doctoral candidate, Karl Hiemenz, to whom he gave the task of constructing a water channel in which he could observe the separation of the flow behind a cylinder. The object was to check experimentally the separation point calculated by means of the boundary-layer theory. For this purpose, it was first necessary to know the pressure distribution around the cylinder in a steady flow. Much to his surprise, Hiemenz found that the flow in his channel oscillated violently. When he reported this to Prandtl, the latter told him: "Obviously your cylinder is not circula". However, even after very careful machining of the cylinder, the flow continued to oscillate. Then Hiemenz was told that possibly the channel was not symmetric, and he started to adjust it. I was not concerned with this problem, but every morning when I came in the laboratory I asked him, "Herr Hiemenz, is the flow steady now?" He answered very sadly,"It always oscillates." "von Kármán thought that this phenomena was partly intrinsic: he analysed the linear stability of point vortex configurations and established a link between the drag on the cylinder and the structure of the vortex street. Later (1954), he wrote "The arrangement of the vortices is connected with my name: it is usually called a Kármán vortex street. But I do not claim to have discovered these vortices: they were known long before I was born. The earliest picture in which I have seen them is one in a church in Bologna, Italy, where St Christopher is shown carrying the child Jesus across a flowing stream. Behind the saint's naked foot the painter indicated alternating vortices"[13]. In 1908, Henri Bénard reported that "the ratio of the wavelength

Fig. 10.3. Visualization of the vortex behind a ciruclar cylinder (Courtesy of T. Leweke).

to the diameter λ/D is independent of the velocity" (cf. Figure 10.3). In 1911 and 1912, von Kármán [14, 15, 16] predicted the stability of an alternate double street of point vortex for a precise value of the geometrical ratio $h/\lambda = 0.28$. Bénard performed many experiments to check this value and to specify the variation of the frequency (*Sur les lois de la fréquence des tourbillons alternés détachés derrière un obstacle*, 7 juin 1926 [7]). "The ratio h/λ varies from 0.15 to 0.49 with an average value around 0.32" (*Sur l'inexactitude, pour les liquides réels, des lois théoriques de Kármán relatives à la stabilité des tourbillons alternés,*

On the inaccuracy, for real liquids, of the theoretical laws of Kármán relative to the stability of alternate vortices, 21 juin 1926 [8]). He has also corrected his previous statement of 1908: "λ/D varies slightly with the velocity V and the Reynolds number" and "when the Reynolds number goes from 73 to 1202, St the Strouhal parameter has changed from 0.092 to 0.25,..." (*Sur les écarts des valeurs de la fréquence des tourbillons alternés par rapport à la loi de similitude dynamique*, 5 juillet 1927 [9]). In fact, on this subject also, we can recall the name of Lord Rayleigh [17, 18], who proposed expressing the Strouhal number as a development of the different powers of $1/Re$ and was interested in the emission of sound associated with the vortex shedding process. According to von Kármán, C. Runge was the first to make the connection between these phenomena.

10.2 The Instability Threshold and the Landau Model

Since the time of Bénard and von Kármán, many hundreds of papers have been published on this instability. This very large number is linked to the importance of applications in different engineering fields and to the apparent simplicity of this configuration which seems to be two-dimensional. In this review, we mainly focus on phenomena observed in the Reynolds number range $Re \in [20, 300]$ where there has been much progress during the last decade in our understanding of three-dimensional phenomena and of the transition to turbulence. Even, if many engineering problems appear for higher Reynolds numbers, it has been shown that the study of the first bifurcations leading to oblique or parallel shedding and its control are valuable for higher values. We refer to the reviews of Berger and Wille [19], Coutanceau and Defaye [20], [21] and to the book of Blevins [22] and Zradkovich[23] which give a description of the different regimes on a wider Reynolds number range. Von Kármán's analysis bears on an ideal two-dimensional vortex street. The discussion of the criteria concerning the geometrical ratio h/λ has been renewed by taking into account the stabilizing effect of finite core size for a real vortex [24, 25] and three-dimensional instability. However, this analysis does not link the details of the flow to the vortex shedding. The concepts of absolute and convective instabilities have been introduced in plasma physics [26] and applied recently to shear flows, see the review of Huerre and Monkewitz [27]. They explain the development of a global mode of wake oscillations by the existence of regions of absolute and convective instability behind the body, due to the downstream evolution of the transverse velocity profile. The selection of the frequency is linked to the feedback of a perturbation upstream and downstream when the size of the region of absolute instability is wide enough to create a global mode. The nonlinear effects and the saturation of this mode are well described by the Landau model.

In his original paper, E. Hopf [28] underlined that the bifurcation of a periodic solution from a stationary solution is observable for example, in the flow around a solid body. The approach developed by Landau [29] analyzes the stability of a steady flow. A nonstationary perturbation of the steady solution of the Navier–Stokes equation is expanded as a sum of modes $A(t)$. Let us consider a mode

$A(t)$ proportional to $e^{\sigma_r t}$ with relative growth rate:

$$\sigma = \sigma_r + i\sigma_i = \sigma_r(1 + i\,c_0) \tag{10.1}$$

In the subcritical regime, $Re < Re_c$, all disturbances are stable and σ_r is negative. At the instability threshold, when $Re = Re_c$, one normal mode is marginally stable, $\sigma_r = 0$, with two angular frequencies σ_i and $-\sigma_i$. As the Reynolds number is increased above the critical value, the coefficient σ_r becomes positive and, near the instability threshold, it might be considered as a linear variation of the Reynolds number, i.e. proportional to $(Re - Re_c)$. The expression for the amplitude is no longer valid for long times and the Landau model takes into account this nonlinear saturation in a new expression of the time derivative:

$$\frac{dA}{dt} = \sigma\,A - l\,|A|^2\,A, \tag{10.2}$$

where the complex nonlinear term is $l = l_r + i\,l_i = l_r\,(1 + i\,c_2)$. Setting the complex amplitude $A = Me^{i\phi}$, one obtains evolution equations for the modulus M and the phase ϕ:

$$\frac{dM}{dt} = \sigma_r\,M - l_r\,M^3 \tag{10.3}$$

and

$$\frac{d\phi}{dt} = \sigma_i - l_i\,M^2 \tag{10.4}$$

Both the scale of amplitude, $(\sigma_r/l_r)^{1/2}$, and the characteristic transient time of oscillation, $1/\sigma_r$, have been deduced from experiments. The evolution equation of the modulus predicts the variation of the saturated energy of oscillation:

$$M_{eq}^2 = \frac{\sigma_r}{l_r} \quad \propto \quad (Re - Re_c) \tag{10.5}$$

Although the measurements of the mean velocity profile and of the periodic fluctuations have been performed by Kovasznay [30] or Roshko [31], these nonlinear effects were analysed only around 1980, following the results obtained for Rayleigh–Bénard convection (see Chapter 3 by Manneville in this volume). The linear relationship (10.5) provides a very accurate determination of the critical Reynolds number Re_c by extrapolation to zero amplitude, which is now taken equal to 47-50 rather than 40. The absence of any hysteresis, except in the case of experiments with soap films [32], is characteristic of a supercritical bifurcation. In principle, this behaviour should be restricted to a narrow range of Reynolds numbers near the bifurcation threshold, where the Landau approximation is valid. However, for the flow behind circular cylinders and rings [32, 33, 34, 35, 36], the Landau model has been validated from local measurements of velocity fluctuations (Figure 10.4) in a surprisingly large interval of the control parameter $Re \in [50, 150]$. Thus, the distinction, based on visualisations, between two regimes with sinusoidal streakline and vortex formation around Reynolds number 70 [20, 37] fades in a single periodic regime covering a large Reynolds number range. Analysis of transient records (or impulse response) and

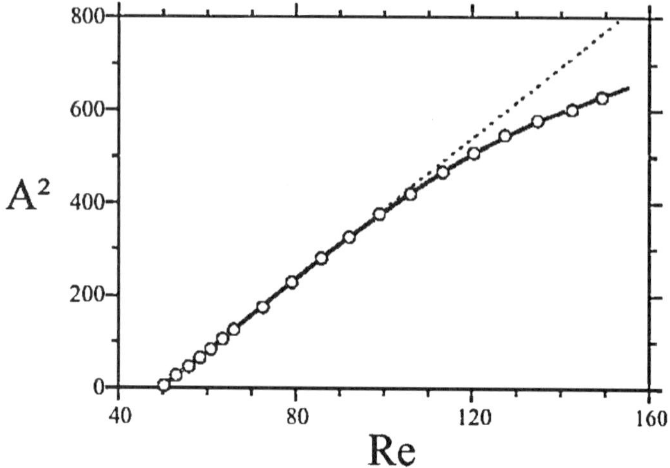

Fig. 10.4. Variation of the streamwise velocity fluctuation with the Reynolds number in the wake of a ring [33].

measurements of the amplitude and frequency have allowed us to determine the variation of the real growth rate with the Reynolds number [12, 39, 40, 42]: $\sigma_r = (Re - Re_c)/5\,d^2$ The coefficient $c_2 = l_i/l_r = -3.0$ has been determined from the nonlinear influence of the amplitude upon the frequency of oscillation [40, 42, 43].

Before the introduction of the Landau model, the Van der Pol oscillator was used by Gaster [44], Blevins [22], and Noack, Ohle and Eckelmann [45] in a qualitative manner. The Landau equation can be deduced as the time-average approximation of a Duffing–van der Pol oscillator [46]. The amplitude of oscillation, i.e. of the global mode, varies in the streamwise direction. Starting from a zero value on the wall of the cylinder, it increases linearly, reaches a maximum where saturation due to nonlinear effects is dominant, and decreases further downstream (see Chapter 11 by Goujon-Durand and Wesfreid in this volume) with the evolution of the velocity profile. The two rows of the vortex street have also been taken as two non-linear coupled oscillators by Pomeau [47], who inquired into the different symmetries of this configuration. Villermaux [48] has proposed a new delay term $A(t - \tau)$ in the Landau equation, which takes into account the convection time $\tau = D/U$ and the influence of the shear layer. Moreover, the experimental relationship $St - Re$ has been recovered on a large Reynolds number range from the threshold to 4000.

The dynamics observed in the wake of axisymmetric objects (spheres, cones, disks) has recently been analyzed in the same way [49, 50]. At low Reynolds number, the flow behind a sphere is axisymmetrical. Above a critical Reynolds number $Re_1 = 212$, the bubble formed downstream of the sphere is no longer axisymmetrical and two trails appear downstream the bubble. The three-dimensional simulations of the birth of the azimuthal mode have been successfully compared to the Landau model by Thompson et al. [51, 52], who have deduced both the

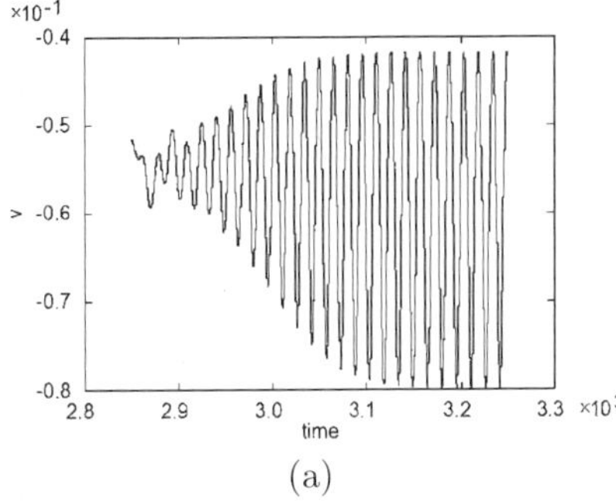
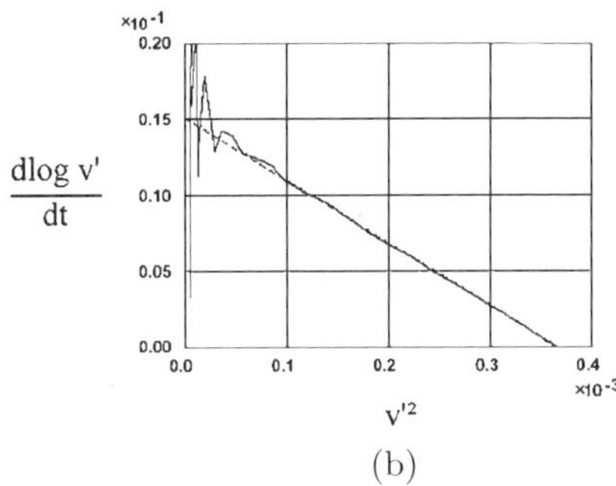

Fig. 10.5. (a) Growth and (b) saturation of the sphere wake instability at $Re = 280$ [51, 52].

linear and nonlinear terms. These two trails oscillate above a second critical Reynolds number Re_2 and for larger values in Reynolds number vortex loops, are periodically shed. In these three-dimensional flows, the spatial axisymmetry break occurs before the transition from steady to periodic flow. A small upstream velocity gradient allows us to control the orientation of the wake. Although small extrinsic perturbations might slightly change the critical Reynolds number, the Landau model (Figure 10.5) has been verified in every configuration, in agreement with numerical simulations [51, 52, 53]. Narrow ranges in Reynolds number above the oscillation threshold were found, where the wakes are periodic and where Landau's law is valid. Downstream of the sphere, the maximum of oscillation has been measured and the variation of the renormalized energy of oscillation is similar to the universal curve described in the cylinder configuration. However, in the case of the sphere, the variation of the location of the maximum changes from six diameters near the threshold to four diameters at the upper value of the periodic range. This behavior is quite different from the wake of a cylinder where the location of the maximum becomes infinite as $(Re - Re_c)^{-1/2}$ near the threshold.

10.3 Three-Dimensional Patterns and the Ginzburg–Landau Model

In some experiments, Henri Bénard noticed the irregular shedding of vortices and we can now see these phenomena as the consequence of three-dimensional effects. A number of studies in recent years have shown the importance of three-dimensional phenomena in the wake of nominally two-dimensional bodies, mainly circular cylinders, and have pointed out the importance of end effects in creating some of the three-dimensionalities [21]. Various patterns appear such as oblique vortex shedding [19, 54], cells of different frequencies [55], "chevrons" [56], and vortex dislocations. For more than a century, the relationship between the frequency of vortex shedding and the velocity of the flow has been the subject of polemics. Thus, the Strouhal–Reynolds number dependence (see Figure 6a from [57]) exhibited a scattering of 20 %, even if all the parameters have been measured with a precision of 1%. The discontinuities in this St-Re curve have been observed by Tritton [58, 59] and discussed by Gaster [44, 60] and by Gerrard [57]. The existence of oblique and parallel vortex shedding was shown in papers of Berger and Wille [19] and of Ramberg [54], but it is only with the paper of Williamson [61] that a nice collapse of the previous data appear by changing the measured value of the Strouhal number S_θ to a new value $St = S_\theta/cos\theta$, after correction of the oblique vortex shedding angle.

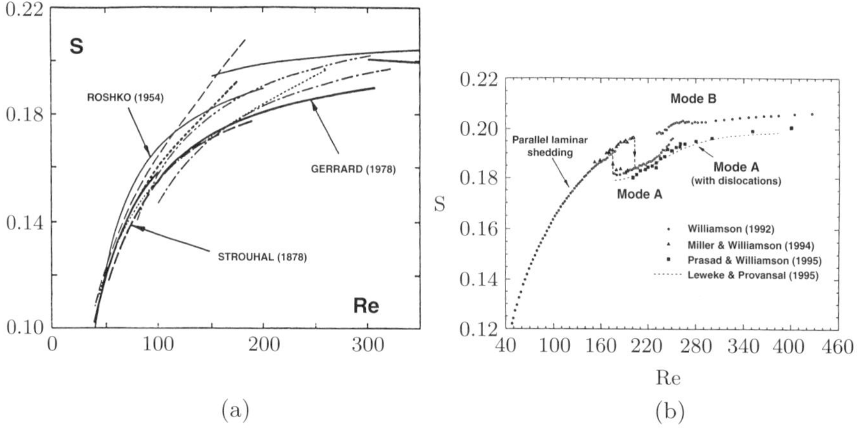

Fig. 10.6. Strouhal–Reynolds variation: (a) From Gerrard [57] and (b) from Williamson [21].

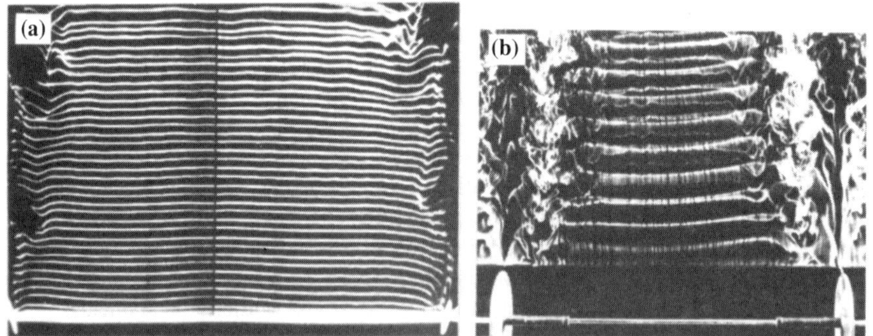

Fig. 10.7. Parallel vortex shedding behind a circular cylinder: (a) From Williamson [56] and (b) with end cylinders [62].

Moreover, different groups have found a parallel vortex shedding by passive control of the flow through angling end plates (Williamson [56] Figure 10.7a), ending larger coaxial cylinders (Figure 10. 7b [62, 63, 64]), large upstream cylinders [65]. The configuration of the ring, studied by Roshko [31], has been investigated by Leweke and Provansal [33, 66] to avoid the end effects and observe a parallel vortex shedding as a natural mode. A universal curve Strouhal number versus the Reynolds number (Figure 10.6b) has been obtained with an accuracy better than 1% between the measurements of different authors [33, 61, 65, 67]. These three-dimensional structures have been described by an extension of the Landau model, including one spatial dimension (the spanwise direction) in which the three-dimensional wake is now represented by a one-dimensional chain of nonlinear coupled oscillators. A thin slice of the wake (for example between z and $z + dz$ is a local oscillator governed by the Landau equation. A diffusive coupling along the spanwise direction is added with a complex diffusive coefficient $\mu = \mu_r + i\mu_i = \mu_r(1 + i\,c_1)$. The velocity fluctuations in the near–wake are assumed to be proportional to the real part of a complex amplitude, solution of the one-dimensional complex Ginzburg–Landau equation with appropriate boundary conditions:

$$\frac{\partial A}{\partial t} = \sigma A - l\,|A|^2\,A + \mu\,\frac{\partial^2 A}{\partial z^2} \tag{10.6}$$

The introduction of the coefficient μ_r leads to a new scaling along the spanwise direction $(\mu_r/\sigma_r)^{1/2}$ which goes to infinity near the instability threshold. A reduced length has been introduced [42]:

$$L_{red} = L\left(\frac{\sigma_r}{\mu_r}\right)^{1/2} = \left(\frac{\nu}{\mu_r}(Re - Re_{c0})\right)^{1/2}\frac{L}{d}, \tag{10.7}$$

where Re_{c0} is the critical Reynolds number for the parallel vortex shedding of an infinite cylinder ($L/d \rightarrow \infty$ $Re_{c0} = 47 - 50$). The linear theory predicts the instability of a spanwise mode n when the reduced length is greater than $n\pi$. This approach results in a prediction for the variation of the critical Reynolds number with aspect ratio, which is in good agreement with experimental measurements. This simple model shows that both the aspect ratio with boundary conditions and the Reynolds number are control parameters of the instability of real three-dimensional flows. The control of the wake by a local retroaction creates a hole or zero amplitude of oscillation that acts as a new boundary condition where the wake can be described by a Ginzburg–Landau equation [68]. By increasing the reduced length, more and more spatial modes appear: first the single mode regime, then the second mode [42] and for larger values chevrons (see Figure 10.8 (a) chevrons [56] and (b) simulation by Albarède [46]), which are created by the symmetrical shock of two oblique waves starting from the boundaries. The chevrons give place to oblique vortex shedding when the boundary conditions are strongly different. The presence of a small velocity gradient near the ends of the cylinder leads to the existence of cells of different frequencies and dislocations, obvious on an isophase simulation graph and similar to the experimental details.

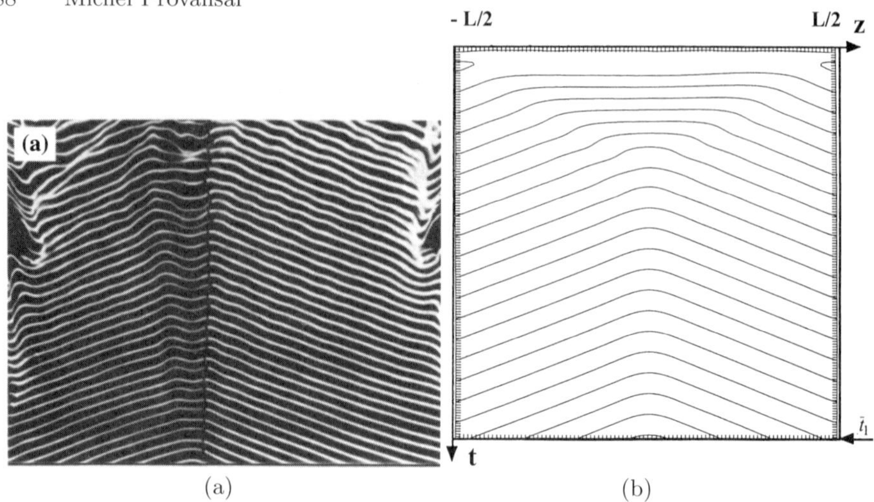

(a)   (b)

Fig. 10.8. (a) Visualization of chevrons by Williamson [56], and (b) simulation from [42].

As for the Landau mode, the parameter in this equation $\mu_r = 10\,\nu$ and c_1 have been determined from experiments [33, 42], because no formal derivation from the equations of motion is available so far. The symmetry $z, -z$ is preserved by this modeling. However this model has also been successful in describing the cells of different frequencies behind a cone [69, 70] or behind a cylinder with an upstream velocity gradient [71] in configurations where the symmetry along the spanwise direction is broken and where additional terms in $\partial A/\partial z$ should be introduced. On short distances in the spanwise direction, which is the size of the different cells, the diffusive term is also sufficient. Facchinetti et al. [72] have analyzed these three-dimensional effects through a van der Pol equation and a coupling along the spanwise direction. A more complete modeling has been examined by Chiffaudel [73] working on a Ginzburg–Landau equation with spatial derivatives in both the spanwise and the streamwise directions to reflect the role played by the downstream evolution in the mechanism of the absolute instability. On the ring configuration, different oblique modes or helices can coexist for the same Reynolds number (Figure 10.10) [33]. Their conditions of stability have been experimentally studied and compared to the predictions of the Ginzburg–Landau model with periodic conditions along the spanwise direction. The transition from oblique unstable mode to parallel stable mode was the first observation of Eckhaus instability in open flows. After the ring, the propagation of phase shocks and phase expansion have been observed in transient patterns of cylinder wake by the manipulation of end effects [74, 75].

10.4 The Transition to Turbulence: The Modes A and B

Because the three-dimensional effects in the laminar regime, such as cells, dislocations, oblique and parallel vortex shedding, have been recently clarified, the

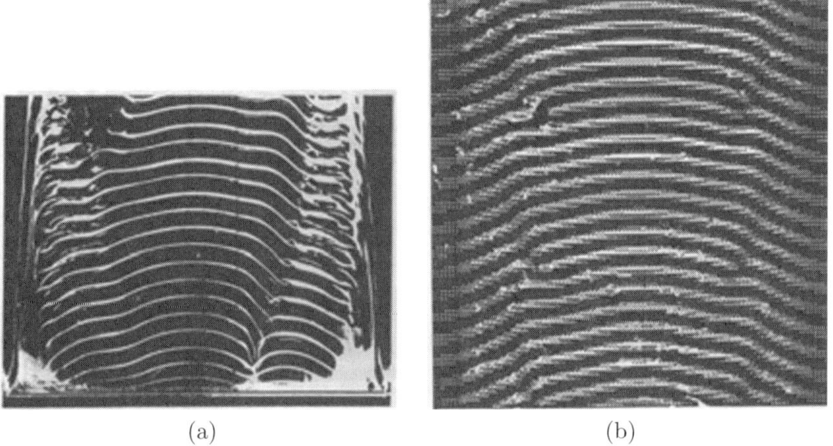

(a)   (b)

Fig. 10.9. (a) Cells of different frequencies [56] and (b) simulation by the Ginzburg–Landau equation [42].

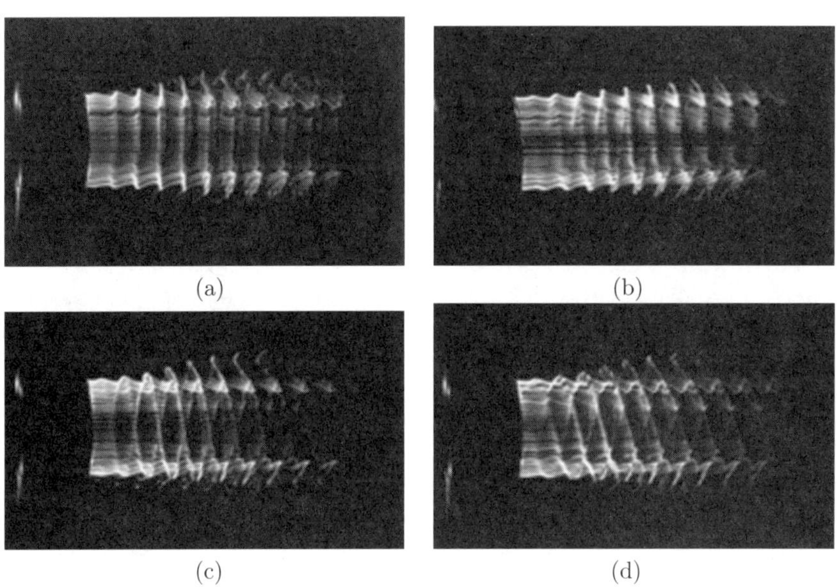

(a)   (b)
(c)   (d)

Fig. 10.10. Parallel and oblique modes in the wake of a ring: (a)Parallel mode, (b) helix mode 1, (c) helix mode 2, and (d) helix mode 3 [33].

transition regime has received a renewed attention during the last decade. The transition to three-dimensionality of the two-dimensional wake of a circular cylinder happens in two distinct stages that can be described with reference to the measurements of the Strouhal–Reynolds curve (Figure 10.6b and Williamson [61]). Two discontinuous changes are visible. The first is in the Reynolds number range $Re \in [150, 190]$, where the spanwise wavelength of this large scale mode A is around four diameters. Increasing the Reynolds number up to 230-260, there is again a discontinuity and a new mode B with spanwise modulation along the fine scale of one diameter. These two modes A and B have distinct unstable wavelength bands and different topologies and they lead rapidly to a fully turbulent flow for higher Reynolds number [21, 76].

There is a wide range $Re \in [150, 190]$ of experimental critical Reynolds numbers for wake transition reported in the literature. The level of the free-stream turbulence has been proposed by Bloor [77] as a cause of this dispersion, although the end conditions are unknown in these experiments. Miller and Williamson [78] have found that the laminar regime for parallel shedding with clean end conditions can be extended up to a critical value $Re_c = 194$ and even beyond $Re = 200$ for short periods of time. Experiments as well as numerical simulations have confirmed the hystereretic character of this bifurcation. In fact, the subcritical character of that transition explains the dispersion upon the critical value of each threshold between different experiments, depending on the turbulence level and whether the flow speed is increased or decreased. Floquet stability analysis [79, 80] predicted that mode A first becomes unstable for a spanwise wavelength $\lambda = 4D$ at a critical value of the first bifurcation, $Re_c = 189$, pretty close to the more recent and well-controlled experimental data. This is consistent with the experimental flow visualizations of Williamson [61] which show the spanwise wavelength to be between $3D$ and $4D$ or 3/5 to 4/5 primary wavelengths (the injection of polymers allowed Cadot and Kumar [81], to change the primary wavelength while keeping the same ratio 4/5 for the spanwise wavelength of the mode A). In this mode, the primary vortices deform in a wavy fashion during the shedding process (Figure 10.11a). This results in the formation of vortex loops stretched into streamwise vortex pairs. The mode A structure and its length scale are similar to the "in-phase" mode of vortex loop formation in an unseparated wake behind a splitter plate studied by Meiburg and Lasheras in 1988 [82]. There are also strong indications that the equivalent of mode A for a circular cylinder is the initial transition mode for various two-dimensional bodies, such square cylinders [83] and long plates with aerodynamic noses [84]. At higher Reynolds numbers, finer-scale streamwise vortex pairs are formed as shown in Figure 10.11b. The primary vortex deformation is more spanwise-uniform than for mode A and the streamwise vortex structure has a smaller spanwise wavelength of around 1/5 of a primary wavelength. Unlike mode A, mode B appears unstable over a small wavelength band, even at higher Reynolds numbers. Some characteristic features of mode B have been observed through visualizations and spanwise cross-correlations at $Re = 1000$ [85]. For the cylinder as for the ring configuration, the spectra of velocity fluctuations exhibit the same change from

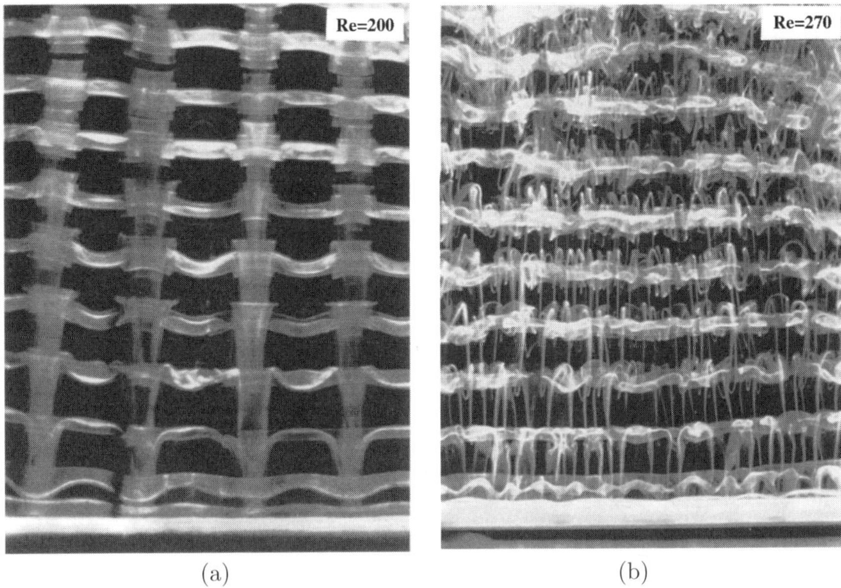

(a)   (b)

Fig. 10.11. Visualizations of the spatial structure of modes: (a) A and (b) B [21].

one single narrow frequency to twin peaks of the larger spectrum and again a single narrow peak when gradually transferring the energy from mode A to mode B in the transition regime. At $Re = 260$, Floquet analysis shows that the two-dimensional wake becomes unstable to a second shedding mode and the critical wavelength $\lambda = 0.8D$ is again consistent with experimental observations. Thompson, Hourigan, and Sheridan [86] used a spectral element/Fourier series discretization and they captured computationally the modes A and B for the first time (Figure 10.12 a,b). The natural symmetries of modes A and B are different: mode A exhibits a staggered sequence of streamwise vortices from one braid to the next one and mode B comprises an inline arrangement ([21], Figure 10.13). Lasheras and Meiburg [87] have found two types of modes induced by initial perturbations that have the symmetries of modes A and B respectively. Despite the large number of experimental, theoretical, and numerical studies of this transition, the precise nature of the secondary instabilities is still not fully understood. Barkley et al. [88] showed that the development of three-dimensional flow in the wake of a circular cylinder, including the hysteretic onset of mode A and the energy shift from mode A to mode B, could be described by a coupled pair of evolution equations. Similar results have been also obtained by Sheard et al. [34] from numerical simulations. Williamson [21] realized that the two different instabilities are associated with the two different length scales of the basic two-dimensional flow. These scales are the core sizes of the primary vortices and the width of the braids between the rollers. Leweke and Williamson [89] showed that the elliptic instability theory of the vortex cores predicts the spanwise wavelength of the mode A instability and explains the topology and the waviness of

(a)   (b)

Fig. 10.12. Numerical simulation of Modes (a) A and (b) B (from Thompson et al. [86]).

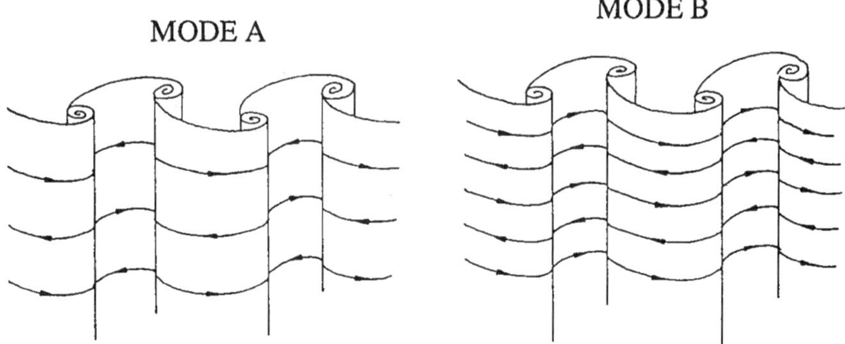

Fig. 10.13. Sketches of the symmetries of Modes A and B (from Williamson [21]).

the core vortices. Henderson [76] was critical of this scenario mainly because the mode appears to have the largest amplitude outside the region in which instability theory predicts it should develop. Direct numerical simulations (see, for instance, [51, 52, 90, 91]) taking into account the contributions of elliptic and hyperbolic flow regions to the three-dimensional transition, allow us to conclude that the elliptic instability is dominant in the initiation and maintenance of mode A perturbation. Mode B instability has been associated by Williamson with an instability of the braid region and more precisely scales on the thickness of the vorticity layer lying in the braid region.

10.5 Diversity of Applications

The work of Strouhal and Bénard has strong applications in aeroacoustics [23]. The destruction of Tacoma Bridge under the action of the wind occurred between the two World Wars and many studies have been devoted to the knowledge of the vortex shedding frequency to avoid resonance of the structure. Thus, in France at Ecole Normale Supérieure, the team of Yves Rocard [92] was involved

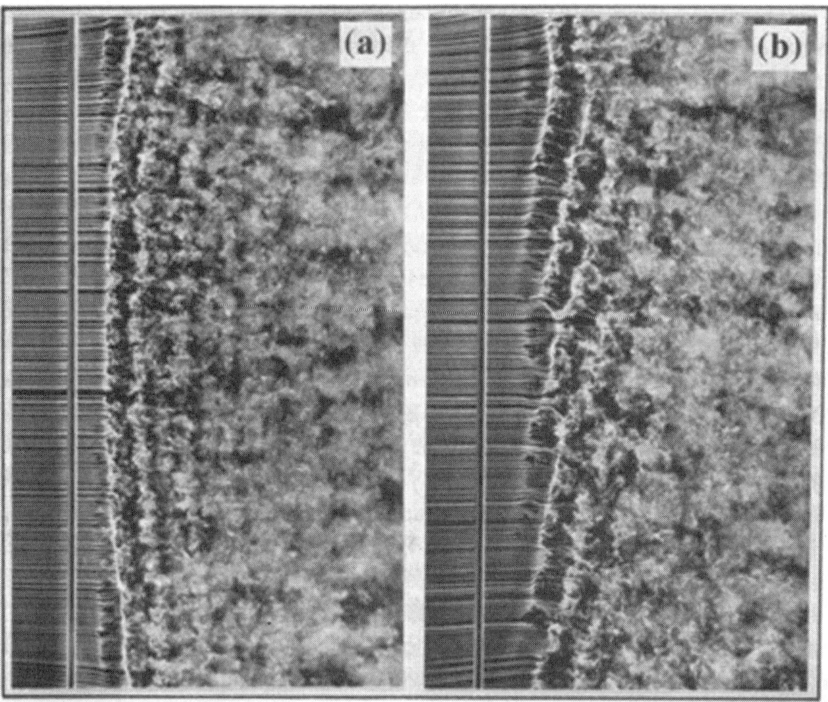

Fig. 10.14. (a) Parallel and (b) oblique vortex shedding for $Re = 10^4$ [21].

in the conception of the Tancarville Bridge near Le Havre. The variation of the frequency with the flow velocity led to the conception of a new class of flow-rate instruments. The detection of periodic velocity fluctuations by electronic system avoids the rotation of mechanical systems and different patents have been developed. Usually, tapered bluff bodies are selected rather than circular cylinder shapes to get the vortex birth from the same point of the solid surface.

Moreover, the wake of a circular cylinder appears also as a prototype of the wake behind a bluff body (boat, car, plane). The Album of Flow Motion [93] shows the vortex shedding of petroleum behind a tanker. The vibrations of cables in the wind or under the sea are of great interest in different fields: telecommunications, offshore petroleum industry (cf. the review of Bearman [94], and the three Bluff Body Wakes and Vortex Induced Vibrations Conferences B.B.W.V.I.V 1,2,3 1998, 2000 and 2002 on these problems [95]). The use of a sinuous cylinder along the spanwise direction [96], instead of a straight one, leads to the reduction of amplitude and even to the suppression of vortex shedding. In the previous section, we have related the diversity of three-dimensional phenomena at a low Reynolds number. The methods used to promote parallel vortex shedding or oblique vortex shedding are still useful for $Re = 10^4$, as it is visible on the Figure 12 of Williamson [21]. Parallel vortex shedding results in a collective synchronised effect of forces working on the whole span of the cylinder, when oblique vortex shedding will induce a weaker force on a shorter span interval. The feedback control of the flow by an actuator can suppress the vortex shedding [68].

The general properties of the wake are important also in heat transfer and chemical engineering. Many empirical correlation laws used in heat transfer are linked to the different regimes of oscillation. The influence of the periodic shedding on the temperature distribution is obvious on the Figure 15 in Mathelin, Bataille, and Lallemand's study [97] which was motivated by the blowing of cool gas in a turbomachine. We can also notice that the transverse distance between the two rows is weak; in this case the wake is sometimes modeled as a single row of alternate vortices. In the tube bank of heat exchangers [22], or around

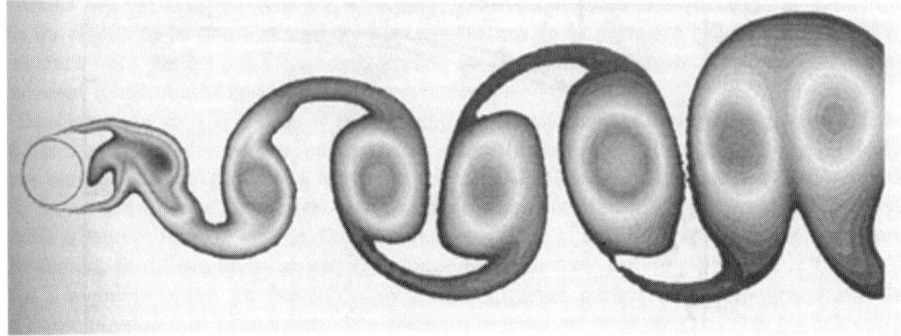

Fig. 10.15. Temperature field behind a heated cylinder [96].

the columns of some bridges, the wakes behind bluff bodies placed next to each other can interact and create a large variety of phenomena. An extension of the Landau model has been proposed to study N coupled wakes which showed that the wakes of a couple of cylinders [97] are uncoupled when the distance separating their axes is larger than six times their diameter. For a larger number of wakes, the coupled oscillators system takes the form of a discrete version of the Ginzburg–Landau equation, where the coupling parameter is a function of the distances separating the cylinder axes. At strong coupling, the entire population of oscillators move in phase, at intermediate coupling spatio-temporal chaos occurs and finally at weak coupling the wakes oscillate in phase opposition. These theoretical predictions have been verified on a row of 21 identical cylinders placed perpendicular to a uniform flow and separated by a distance equal to 1.5 times the cylinders diameter, in order to create a very strong coupling. The size of the cells depends on the intial conditions of the flow. Sligthly above $Re = 110$, some vortex streets are generated by the cylinders belonging to some of the clusters. These oscillating wakes, gathered in some regions of the row, are locked in phase. These groups exist together with stationary regions, creating amazing patterns (Figure 10.16). A decrease of the coupling by changing the distance between neighbours shows the existence of the phase opposition mode. Figure 10.17 illustrates the permanence of these structures in different configurations.

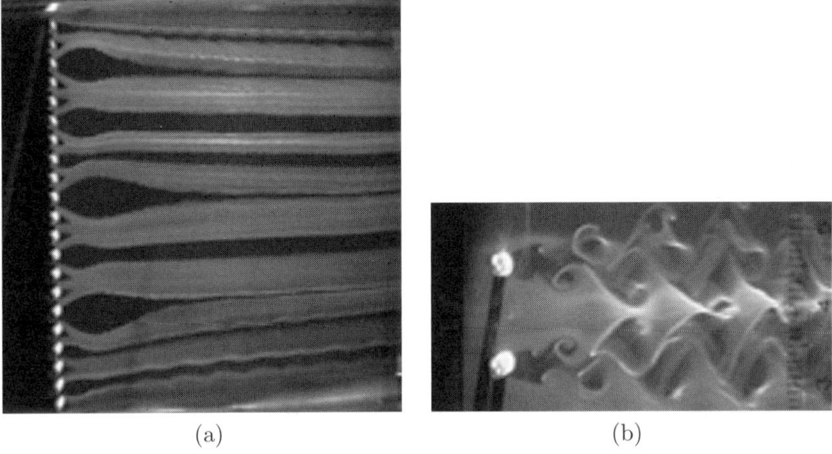

(a)   (b)

Fig. 10.16. Coupled wakes of (a) a row and (b) a pair of circular cylinders [98].

The spatial photographs show the existence of regular alternate vortex shedding behind mountains or volcanoes (Jan Mayen in the North Atlantic, Guadalupe, or Socorro Island in Gulf of California or Cheju in the south of Korea, Figure 10.17-a), even at very high Reynolds number and in stratified flows. Couder, Chomaz, and Rabaud [100] have used soap films to study two-dimensional cylinder wakes. This technique has been developed in another way by Gharib and Derango [101] by a flow of liquid along wires. Even if some aspects of these dynamics and the role of the film thickness might be clarified, the study of the wake of a flexible filament has been undertaken by Libchaber's team [102] (Figure 10.18a,b,c). This flow is one kind of prototype of one-dimensional flag instability. Moreover, number of applications appear in biology such the swimming of small organisms or the flight of insects.

10.6 Conclusion

Since the times of Bénard, much progress has been made in experimental techniques, three-dimensional simulations, and theoretical scenarios of this instability. However, except for the explanation of convective and absolute instability, a theory of the linear threshold of this instability is still lacking. A number of qualitative and quantitative features of bluff-body wakes can be interpreted using relatively simple amplitude equations. Different configurations such as cylinders, rings, and axisymmetrical objects have been analyzed in this way. During the last decade, the two-dimensional flow behind a circular cylinder has been documented thanks to a careful control of end conditions. Three-dimensional instabilities have been observed and characterized through visualizations and quantitative data. The measurements of the amplitude and frequency of oscillations, and wavelengths of primary instability have been obtained on different

Fig. 10.17. Left vortex shedding behind the Guadalupe Island [103].

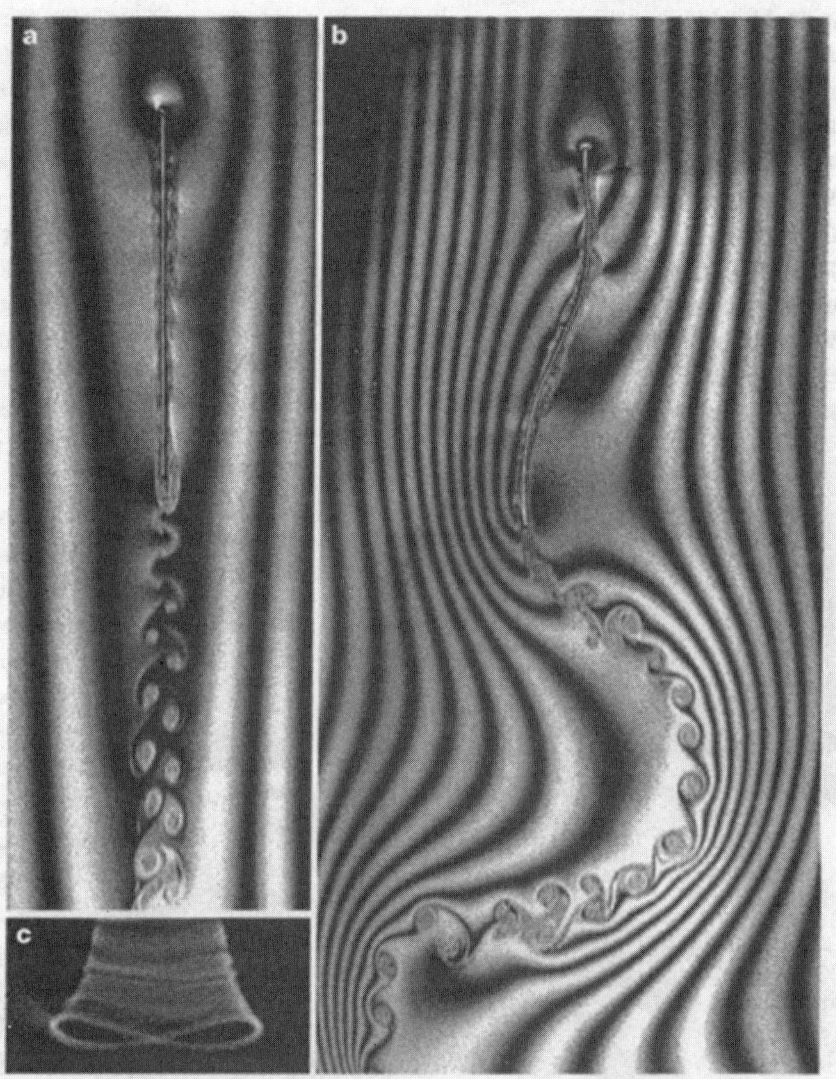

Fig. 10.18. Model of one-dimensional flags in two-dimensional wind [102].

experimental setups with accuracy. The flow bifurcations that occur at various Reynolds numbers have been identified and accurately described. The direct numerical three-dimensional simulations are now a powerful tool to check the elliptic or hyperbolic root of the mechanisms leading to the formation of modes A and B. The approximate spanwise wavelengths of these modes have been computed and successfully compared to the experiments. Various fields of applications find interest in these developments but it is remarkable that, after a very large amount of work motivated by specific applications, the progress of our understanding came about through from basic experiments. In conclusion, the artistic aspect of the patterns observed by Henri Bénard should be underlined and it is still worthwhile to watch these vortices in a river [103] or the painting of Madonna col Bambino tra i Santi Demenico, Pietro Martire e Cristoforo in the Basilica di San Domenico, Bologna, Italy [104].

Acknowledgments

The author would like to thank Thomas Leweke, Patrice Le Gal, and Lionel Schouveiler for help in the preparation of this review. Thanks also to Pierre Albarède, Françoise Bataille, Mark Thompson, and Charles Williamson in providing material for this review.

References

1. V. Strouhal, Uber eine besondere Art der Tonerregung. Wied, *Anna. Phys. und Chem.* (Leipzig) Series **3**, 216–251 (1878).
2. H. Bénard, Formation de centres de giration à l'arrière d'un obstacle en mouvement, 9 novembre 1908, *Compt. Rend. Acad. Sci.* Paris **147**, 839–842 (1908).
3. H. Bénard, Étude cinématographique des remous et des rides produits par la translation d'un obstacle, 23 novembre 1908, *Compt. Rend. Acad. Sci* Paris, **147**, 970–972 (1908).
4. A.Mallock, On the resistance of air, *Proc. Roy. Soc.* **A79**, 262–273 (1907).
5. H. Bénard, Sur la zone de formation des tourbillons alternés derrière un obstacle, 31 mars 1913, *Compt. Rend. Acad. Sci.* Paris (1913).
6. H. Bénard, Sur la marche des tourbillons alternés derrière un obstacle, 21 avril 1913, *Compt. Rend. Acad. Sci.* Paris (1913).
7. H. Bénard, Sur les lois de la fréquence des tourbillons alternés détachés derrière un obstacle, 7 juin 1926, *Compt. Rend. Acad. Sci.* Paris (1926).
8. H. Bénard, Sur l'inexactitude, pour les liquides réels, des lois théoriques de Kármán relatives à la stabilité des tourbillons alternés, 21 juin 1926, *Compt. Rend. Acad. Sci.* Paris (1926).
9. H. Bénard, Sur les écarts des valeurs de la fréquence des tourbillons alternés par rapport à la loi de similitude dynamique, 5 juillet 1927, *Compt. Rend. Acad. Sci.* Paris (1927).
10. D. Riabouchinsky, *L'Aérophile*, **19**, 15 (1911).
11. G. von dem Borne, *Zeits. Für Flugtechnik*, **3** (1912).
12. T. von Kármán, *The Wind and Beyond: Theodore von Kármán Pionneer in Aviation and Pathfinder in Space*, Little Brown, New York (1967).
13. T. von Kármán, *Aerodynamics*, Cornell University Press, Ithaca, New York, see also McGraw-Hill paperback (1954).
14. T. von Kármán, Uber den Mechanismus den Widerstands, den ein bewegter Korper in einer Flussigkeit erfahrt, *Gott. Nachr*, part 1: 509–517 (1911).
15. T. von Kármán, Uber den Mechanismus den Widerstands, den ein bewegter Korper in einer Flussigkeit erfahrt, *Gott. Nachr*, part 2: 547–556 (1912).
16. T. von Kármán and H. Rubach, *Phys. Z* **13**, 49–59 (1912).

17. L. Rayleigh, *Theory of Sound*, second edition (1896).
18. L. Rayleigh, Aeolian tones, *Phil. Mag.* **29**, 433 (1915).
19. E. Berger and R. Wille, Periodic flow phenomena, *Ann. Rev. Flluid. Mech.* **4**, 313 (1972).
20. M. Coutanceau and J.-R. Defaye, Circular cylinder wake configurations: a flow visualization survey, *Appl. Mech. Rev.* **44**, 255–305 (1991).
21. C.H. K. Williamson, Vortex Dynamics in the cylinder wake, *Ann. Rev. Fluid Mech.* **28**, 477–539 (1996).
22. R.D. Blevins, *Flow-Induced Vibrations*, New York: Van Nostrand Reinhold (1990).
23. M. Zdravkovich, *Flow Around Circular Cylinders*, Oxford University Press, Oxford (2002).
24. J. Jimenez, On the linear stability of the inviscid Karman vortex strcct, *J. Fluid Mech.* **178**, 177–194 (1987).
25. P.G. Saffman, *Vortex Dynamics*, Cambridge University Press (1992).
26. A. Bers, Space time evolution of plasma instability absolute and convective, in *Handbook of Plasma Physics* (ed. M.N. Rosenbluth and R.Z. Sagdeev), 1, 451–517 (1983).
27. P. Huerre and P.A. Monkewitz, Local and global instabilities in spatially developing flows, *Ann. Rev. Fluid Mech.* **22**, 473–537 (1990).
28. E. Hopf, Abzweigung einer periodischen Losung von einer station aren Losung eines Differentialsystems, *Ber. Ver h. S achs. Akad. Wiss.* Leipzig, Math.-phys. Kl., **94**, 1–22 (1942). Translated in *The Hopf Bifurcation and Its Applications* (eds. J.E. Marsden and M. McCracken), Springer, New York, pp. 163–193 (1976).
29. L.D. Landau, On the problem of turbulence, *C. R. Acad. Sci.* URSS **44**, 311–314 (1944).
30. L. Kovasznay, Hot wire investigation of the wake behind cylinder at low Reynolds numbers, *Proc. Roy. Soc.* **A**, **198**, 174–189 (1949).
31. A. Roshko, On the development of turbulent wakes from vortex streets, *NACA Report* 1191, National Advisory Committee for Aeronautics, Washington, DC (1954).
32. V.K. Horváth, J Cressman, W.I. Goldburg, and X.L. Wu, Hysteretic transition from laminar ro vortex shedding flow in soap films, *Advances in Turbulence VIII European Turbulence Conference* (Dopazo et al. eds.), (CIMNE, Barcelona 2000).
33. T. Leweke and M. Provansal, The flow behind rings: bluff body wakes without end effects, *J. Fluid Mech.* **288**, 265–310 (1995).
34. G.J. Sheard, M.C. Thompson, and K. Hourigan, A coupled Landau model describing the Strouhal-Reynolds number profile on a three-dimensional circular cylinder wake, *Phys. Fluids* **15**(9), 68–71 (2003).
35. G.J. Sheard, M.C, Thompson, and K. Hourigan, From spheres to circular cylinders: the stability and flow structures of bluff ring wakes, *J. Fluid Mech.* **492**, 147–180 (2003).
36. G.J. Sheard, The stability and characteristics of the flow past rings, Ph.D. dissertation Monash University, Australia (2004).
37. R.B. Green and J.H. Gerrard, An optical interferometric study of the wake of a bluff body, *J. Fluid Mech.* **226**, 219–242 (1991).
38. C. Mathis, M. Provansal, and L. Boyer, The Bénard–von Kármán instability: an experimental study near the threshold, *J. Physique Lett.* **45**, L-483–491 (1984).
39. M. Provansal, C. Mathis, and L. Boyer, Bénard–von Kármán instability: transient and forced regimes, *J. Fluid Mech.* **182**, 1–22 (1987).
40. K.R. Sreenivasan, P.J. Strykowski, and D.J. Olinger, Hopf bifurcation, Landau equation and vortex shedding behind cylinders, in *Forum on Unsteady Flow Separation* (ed. K.N. Ghia), ASME, New York, FED **52**, 1–13 (1986).

41. H. Oertel, Wakes behind blunt bodies, *Ann. Rev. Fluid Mech.* **22**, 539 (1990).

42. P. Albarède and P. A Monkewitz, A model for the formation of oblique shedding and chevron patterns in cylinder wakes, *Phys. Fluids A* **4**,. 744–756 (1992).

43. M. Schumm, E. Berger, and P.A. Monkewitz, Self-excited oscillations in the wake of two-dimensional bluff bodies and their control, *J. Fluid Mech.* **271**, 17–53 (1994).

44. M. Gaster, Vortex shedding from slender cones at low Reynolds numbers, *J. Fluid Mech.* **38**, 565–576 (1969).

45. B.R. Noack, F. Ohle, and H. Eckelmann, On cell formation in vortex streets, *J. Fluid Mech.* **227**, 293–308 (1991).

46. P. Albarède and M. Provansal, Quasi-periodic cylinder wakes and the Ginzburg-Landau model, *J. Fluid Mech.* **291**, 191–222 (1995).

47. Y. Pomeau, Remarks on the bifurcations with symmetry, *Chaos, Solitons and Fractals* **5** (9) 1755–1761 (1995).

48. E. Villermaux, On the Strouhal–Reynolds dependence in the Bénard.-Kármán problem, *IUTAM Symposium on Bluff Body Wakes and Vortex-Induced Vibrations* **2**, Marseille (2000).

49. M. Provansal and D. Ormières, Bifurcation from steady to periodic flow in the wake of a sphere, *C. R. Acad. Sci.* Paris (1998).

50. B. Ghidersa and J. Dusek, Breaking of axisymmetry and onset of unsteadiness in the wake of a sphere, *J. Fluid Mech.* **423**, 33–69 (2000).

51. M.C. Thompson, T. Leweke, and M. Provansal, Kinematics and dynamics of sphere wake transition, *J. Fluids. Struct.* **15**, 575–585 (2001).

52. M.C. Thompson, T. Leweke, and C.H.K. Williamson, The physical mechanism of transition in bluff body wakes, *J. Fluids. Struct.* **15**, 607–616 (2001).

53. A.G. Tomboulides and S.A. Orszag, Numerical investigation of transitional and weak turbulent flow past a sphere, *J. Fluid Mech.* **416**, 45–73 (2000).

54. S.E. Ramberg, The effects of yaw and finite length upon the vortex wakes of stationary and vibrating cylinders, *J. Fluid Mech.* **128**, 81 (1983).

55. D. Gerich and H. Eckelmann, Influence of end plates and free ends on the shedding frequency of circular cylinder, *J. Fluid Mech.* **122**, 109 (1982).

56. C.H.K. Williamson, Oblique and parallel modes of vortex shedding in the wake of a circular cylinder at low Reynolds numbers, *J. Fluid Mech.* **206**, 579–627 (1989).

57. J.H. Gerrard, The wakes of cylindrical bluff bodies at low Reynolds number, *Phil. Trans. R. Soc.* London Ser **A**, **288**, 351 (1978).

58. D.J. Tritton, Experiments on the flow past a circular cylinder at low Reynolds numbers, *J. Fluid Mech.* **6**, 547 (1959).

59. D.J Tritton, A note on vortex streets behind circular cylinders at low Reynolds numbers, *J. Fluid Mech.* **45**, 203 (1971).

60. M. Gaster, Vortex shedding from circular cylinders at low Reynolds numbers, *J. Fluid Mech.* **46**, 565 (1971).

61. C.H.K. Williamson, Defining a universal and continuous Strouhal-Reynolds number relationship for the laminar vortex shedding of a circular cylinder, *Phys. Fluids* **31**, 2742 (1988).

62. H. Eisenlhor and H. Eckelmann, Vortex splitting and its consequences in the vortex street wake of cylinders at low Reynolds numbers, *Phys. Fluids A* **1**, 189 (1989).

63. H. Eckelmann, J.M.R. Graham, P. Huerre, and P.A. Monkewitz (eds), *Proc. I.U.T.A.M. Conference Bluff Body Wake Instabilities*, Berlin, Springer-Verlag (1992).

64. M. König, H. Eisenlohr, and H. Eckelmann, The fine structure in the Strouhal-Reynolds number relationship of the laminar wake of a circular cylinder, *Phys. Fluids* **A 2**,1607–1614 (1990).
65. M. Hammache and M. Gharib, An experimental study of the parallel and oblique vortex shedding from circular cylinders, *J. Fluid Mech.* **232**, 567 (1991).
66. T. Leweke and M. Provansal, Determination of the parameters of the Ginzburg-Landau wake model from experiments on a bluff ring, *Europhys. Lett.***27**, 655–670 (1994).
67. C. Norberg, An experimental investigation of the flow around a circular cylinder: influence of aspect ratio, *J. Fluid Mech.* **258**, 287 (1994).
68. K. Roussopoulos, Feedback control of vortex shedding at low Reynolds numbers, *J.Fluid Mech.* **248**, 267–296 (1993).
69. A. Papangelou, Vortex shedding from slender cones at low Reynolds numbers, *J. Fluid Mech.* **242**, 299–321 (1992).
70. D. S. Park and L. G. Redekopp, A model for pattern selection in wake flows, *Phys. Fluids* **A 4**, 1697–1706 (1992).
71. P.A. Monkewitz, *Communication of BBVIV2*, Marseille (2000).
72. M.L. Fachinetti, E.de Langre, and F. Biolley, Vortex shedding modeling using diffusive van der Pol oscillators, *Compt.Rend. Acad. Sci. Mécanique* Paris (2002).
73. A. Chiffaudel, Non-linear stability analysis of two-dimensional patterns in the wake of a circular cylinder, *Euro-phys. Lett.* **18**, 589–594 (1992).
74. P. A. Monkewitz, C.H.K. Williamson, and G.D. Miller, Phase dynamics of Kármán vortices in cylinder wakes, *Phys. Fluids* **8**, 91–96(1996).
75. T. Leweke, M. Provansal, G.D. Miller, and C.H.K. Williamson, Cell formation in cylinder wakes at low Reynolds number, *Phys. Rev. Lett.* **78**, 1259–1262 (1997).
76. R.D. Henderson, Non linear dynamics and patterns formation in turbulent wake transition, *J. Fluid Mech.* **352**, 65–112 (1997).
77. S. Bloor, The transition to turbulence in the wake of a circular cylinder, *J. Fluid Mech* **19**, 290 (1964).
78. G. Miller and C.H.K. Williamson, Control of three-dimensional phase dynamics in a cylinder wake, *Exp. Fluids.* **18**, 26 (1994).
79. D. Barkley and R.D. Henderson, Three-dimensional Floquet analysis of the wake of a circular cylinder, *J. Fluid Mech.* **322**, 215–241 (1996).
80. R.D. Henderson and D. Barkley, Secondary instability in the wake of a circular cylinder, *Phys. Fluids* **8**, 1683–1685 (1996).
81. O. Cadot and S. Kunar, Experimental characterization of viscoelastic effects on two-and three-dimensional shear instabilities, *J. Fluid Mech* **416**, 151–172 (2000).
82. E. Meiburg and J. Lasheras, Experimental and numerical investigation of the three-dimensional transition in plane wakes, *J. Fluid Mech.***190**, 1(1988).
83. J. Robichaux, S. Balachandar, and S.P. Vanka, Three-dimensional Floquet instability of the wake of a square cylinder, *Phys. Fluids* **11**, 560–578 (1999).
84. K. Hourigan, M.C. Thompson, and B.T.Tan, Self-sustained oscillations in flows around long blunt plates, *J. Fluids Struct.*, **15**, 387–398 (2001).
85. J. Wu, J. Sheridan, M.C. Welsh, K. Hourigan, and M. Thompson, Longitudinal vortex structures in a cylinder wake, *Phys. Fluids* **6**, 2883 (1994).
86. M.C. Thompson, K. Hourigan, and J. Sheridan, Three-dimensional instabilities in the wake of a circular cylinder, *Exp. Ther. Fluid Sci.* **12**, 190–196 (1996).
87. J. Lasheras and E. Meiburg, Three-dimensional vorticity modes in the wake of a flat plate, *Phys. Fluids* **A2**, 371 (1990).

88. D. Barkley, L.S. Tuckermann and M. Golubitzky, Bifurcation theory for three-dimensional flow in the wake of a circular cylinder, *Phys. Rev.* **E. 61**, 5247–5252 (2000).

89. T. Leweke and C.H.K. Williamson, *J. Fluid Mech.* **416**, 151–172 (1998).

90. H. Persillon and M. Braza, Physical analysis of the transition to turbulence in the wake of a circular cylinder by three-dimensional Navier-Stokes simulation, *J. Fluid. Mech.* **365**, 23–89 (1998).

91. R. Mittal and S. Balachandar, Effect of three-dimensionality on the lift and drag of nominally two-dimensional cylinders, *Phys. Fluids* **7**, 1841 (1995).

92. Y. Rocard, *L'instabilité en Mécanique: Automobiles, Avions, Ponts Suspendus*, Masson, Paris (1954).

93. M. Van Dyke, *An Album of Fluid Motion*, Parabolic Press, Inc. (1982).

94. P.W. Bearman, Vortex shedding from oscillating bluff bodies, *Ann. Rev. Fluid. Mech.* **16**, 95 (1984).

95. P.W. Bearman, K. Hourigan, T. Leweke, and C.H.K. Williamson (eds), Bluff Body Wakes and Vortex Induced Vibrations Conferences, B.B.V.I.V. 1, 2, 3: Proceedings of the Conferences of Washington 1998, Marseille 2000 (cf. J. Fluid and Structures 2001, vol. **15** $N°3/4$) and Port-Douglas 2002 (to appear in *Eur. J. Fl. Mech.* 2003).

96. J.C. Owen, A.A. Szewczyk, and P.W. Bearman, Suppression of Kármán Vortex Shedding, *Phys. Fluids* **12** (9), S9 (2000).

97. L. Mathelin, F. Bataille and A. Lallemand, Near wake of a circular cylinder submitted to blowing, *Int. J. Heat. Mass Transfer*, **44**, 3701-3708 (2001).

98. I. Peschard and P. Le Gal, Coupled wakes of cylinders, *Phys. Rev. Lett.* **77** (15), 3122 (1996).

99. P. Le Gal, I. Peschard, M.P. Chauve, and Y. Takeda, Collective behavior of wakes downstream a row of cylinders, *Phys. Fluids* **8**, 2097 (1996).

100. Y. Couder, J.M. Chomaz, and M. Rabaud, On the hydrodynamics of soap films, *Physica* **D 37**, 384–405 (1989).

101. M. Gharib and P. Derango, A liquid film (soap film) tunnel to study two-dimensional laminar and turbulent shear flows, *Physica* **D37**, 406 (1989).

102. J. Zhang, S. Childress, A. Libchaber, and M. Shelley, Flexible filaments in a flowing soap film as a model for one-dimensional flags in two-dimensional wind, *Nature* **408**, 835–838 (2000).

103. J. Apt, M. Helfert, and J. Wilkinson, Orbit: NASA Astronauts photograph the Earth, *National Geographic Society* (1996).

104. T. Mizota, M. Zdravkovich, K.U. Graw, and A. Leder, St Christopher and the vortex, *Nature* **404**, 226 (2000).

11 Spatial Inhomogeneities of Hydrodynamic Instabilities

Sophie Goujon-Durand[1]* and José Eduardo Wesfreid[2]

[1] Physique et Mécanique des Milieux Hétérogènes (PMMH–UMR 7636 CNRS)
Ecole Supérieure de Physique et Chimie Industrielles de Paris (ESPCI)
10, rue Vauquelin, 75005 Paris, France
[2] *Université Paris 12, 61 Av. du Général de Gaulle, 94010 Créteil, France
e-mail :[1] sophie@pmmh.espci.fr, [2] wesfreid@pmmh.espci.fr

We describe the spatial inhomogeneities in hydrodynamic patterns, in the cases of confined systems such as the Rayleigh–Bénard convection and open systems such as as the Bénard–von Karman instability of vortex shedding. From experimental results, we define the typical correlation lengths and their scaling law in each situation.

11.1 Envelope of Rayleigh–Bénard Convection

In the study and modeling of hydrodynamic instabilities, the experiments concerning the thermoconvective Rayleigh–Bénard instability were determinant. This instability, created by heating the bottom of a horizontal layer of fluid, presents a remarkable spatial organization in the shape of convection rolls when the temperature gradient applied on the layer reaches a critical value. Modern studies of this instability have permitted the application of the universality concepts and scaling laws from condensed matter physics, to other areas such as fluid mechanics. Rayleigh–Bénard instability has been the subject of many studies, particularly in non-linear dynamics and deterministic chaos [1], as well as spatio–temporal chaos [2] and developed turbulence [3].

The doctoral thesis of Henri Bénard defended at the Sorbonne in 1901, devoted to thermal convection with a free surface, opened the field of systematically experimental study of instabilities for physicists, using accurate methods of observation, particularly optical methods [4]. This field is known as *"Bénard–Marangoni convection."*

One of the subjects that we review briefly is the evolution of the spatial envelope, that is the maximum amplitude of one of the velocity components of the convective rolls, in Rayleigh–Bénard instability. One of the representative values of the state of the system, near onset of instability, is given by the amplitude $A(x)$ which represents the spatial envelope of the roll structures, of wavenumber q in the $2D$ convection. So, for the stationary case, one velocity component is characterized by $V = A(x)cos(qx + \varphi)$, where x is the position variable in the direction perpendicular to the rolls and φ is the phase perturbation. In that situation the amplitude A is the order parameter of the transition. A real flow is characterized by spatial inhomogeneities: presence of wall structure defects, thickness variations, and so on. The extension of macroscopic perturbations, that is the size of the boundary layers in the " rolls" domain, is estimated by equivalence with the correlation length ξ of the system. The coherence length (or influence length), was determined using laser Doppler velocimetry, by measurement the influence on side boundaries of the cells away from L associated with the conditions $A(0) = A(L) = 0$, on the convective structure [5, 6] (Figure 11.1).

The coherence length is related to the relative distance at the onset of instability $\varepsilon = R/R_c - 1$, where R is the Rayleigh number (nondimensional temperature difference) and R_c its critical value at the onset instability. This length shows critical behavior in $\xi \sim \varepsilon^{-1/2}$, whether beyond or below the onset of instability. Other experiments allowed us to determine the amplitude law $A \sim \varepsilon^{1/2}$ in the cells of the large lateral extension and showed the critical slowing down of the temporary perturbations with characteristic time τ varying as ε^{-1}.

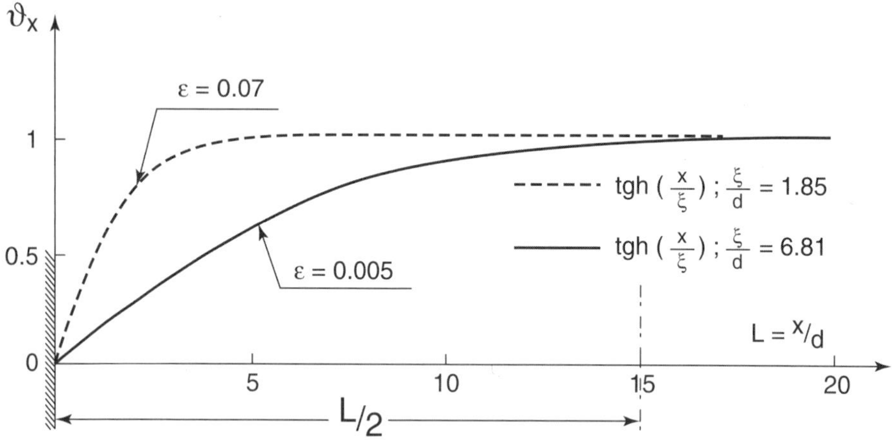

Fig. 11.1. Amplitude modulation induced by wall effects in thermal convection [5].

This series of experiments [7] has permitted, for the first time, the verification of the validity of the slightly non-linear Newell–Whitehead–Segel model which represents the particular case of the Ginzburg–Landau equation for Rayleigh–Bénard convection.

These equations and their generalization have since been widely used as model equations of a whole instabilities problem in confined systems [8]. It is interesting today to recognize that Pierre Bergé and his Saclay group have played a prominent role in the production of these scientific results and in developing optical methods and physical concepts relevant to the Bénard work. The impact of their approach extended to the study of defects in convective structures that have been treated in general terms by considering the effect of compression and deformation of structures at the large scale of the amplitude equation [9, 10, 11].

11.2 Downstream Evolution of the Bénard–von Karman Street

In the last years, theoretical and experimental studies have been focused on hydrodynamic instabilities in open systems. The most typical cases are due to shear forces as in mixing layers or behind a bluff body which induce vortex shedding, in turn provoking transverse vortices in the flow. Other instabilities produce a stationary vortex in the flow direction, a streamwise or longitudinal vortex as a result of centrifugal instabilities due to flow streamline curvature, and the instabilities resulting from rotation or presence of stagnation points. These instabilities are largely represented in real flows either in natural conditions or in industry.

In the first group we have Bénard–von Karman instability, characterized by vortex shedding in the flow behind an obstacle (Figure 11.2). In the case of a cylinder, of infinite lateral extent, the transition takes place when the Reynolds number (nondimensional velocity) calculated on the diameter of the cylinder $R = Vd/\nu$ exceeds the value $R_c = 47$ [12, 13]. The base velocity profile $U(x)$, where x is the distance in the stream direction, presents a strong shearing which is the source of instability. The strongest shearing occurs just behind a bluff body where two recirculation bubbles form. Downstream, the relaxation of the velocity profile, by viscous diffusion, reduces unstable features in the velocity profile. This spatial inhomogeneity of the base flow is characteristic of most of the instabilities in open systems, with, among other consequences, the presence of a normal component of the velocity, through flow conservation. The flow can in this way become non-parallel, if this inhomogeneity is strong enough.

Fig. 11.2. Laser-induced flurosceine view of vortex emission [14].

We are concerned with the spatial inhomogeneities in the Bénard–von Karman instabilities. The bluff body is of a trapezoidal or cylindrical form. We performed experiments in a low velocity water tunnel with a laser Doppler velocimeter. The longitudinal (V_x) and transverse (V_y) velocity components of the instationary flow were measured along the central line of the flow behind the obstacle, for supercritical Reynolds numbers ($R > R_c$), when vortex shedding takes place. In this instability, the vortices behave like traveling waves with wavenumber q and constant frequency f along the flow.

The amplitude or envelope $A(x, y)$, defined in the previous section, accounts for the spatial modulation of intensity of shedding vorticies. Boundary conditions, on the bluff body and to infinity are $A = 0$. In Figure 11.3, we present the envelope $A(x, y_{max})$ of the longitudinal component of the perturbation of flow velocity $v_x = U - V_x$, where y_{max} is the transversal position of the maximum of the perturbation. This figure shows the typical shape of the envelope, increasing the value of maximum and bringing it closer to the bluff body, as the flow velocity or Reynolds numbers R increases [14, 15].

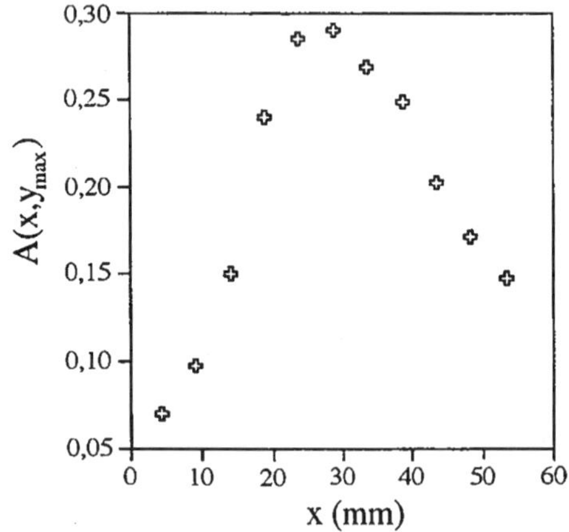

Fig. 11.3. Spatial evolution of the amplitude $A(x, y_{max})$ [15].

As it happens in confined system instabilities, we observe here the deformation of the envelope with the Reynolds number. The value of the amplitude at the maximum of the envelope A_{max} follows a Landau law : $A_{max} \sim \varepsilon 1/2$. The abscise X_{max} of this envelope maximum measured from the extremity of the bluff body, follows the same law as the correlation length in the case of the Rayleigh–Bénard instability, that is $X_{max} \sim \varepsilon -1/2$ [14, 15].

The existence of the two scaling laws allows us to renormalize the envelopes for each value of the Reynolds number, near the onset of instability, by scaling the velocity fluctuation of the velocity with A_{max} and the abscissa with X_{max}. This behavior was also observed in 2D numerical simulations with spectral methods [16] (Figure 11.4). These simulations have also shown that the envelope grows locally in the linear stage of perturbation growth and starts wrapping in the non-linear stage. We obtained similar properties of renormalization in forced wakes [17].

Spatial inhomogeneity of the vortex intensity is correlated with inhomogeneity of the base flow. Local study of the linear stability of the base flow in the region of the near wake close to the recirculation rolls, shows the existence of the absolute instability (IA), that is a natural self-oscillation with narrow spectrum. Downstream the base flow presents a zone of convective instability (IC) with the large noise spectrum amplification [18, 19]. Although each local profile corresponds to instability onset and to different frequencies, the whole unstable system selects one unique frequency characterized by a narrow spectrum (global frequency). The criteria to obtain this global frequency are still subject of theoretical investigation. The unstable flow thus presents a strong synchronization of the oscillations of a global rather than local nature [20]. A retroaction mechanism, common to all flows where an absolute instability region is followed by a convective instability region downstream, results in the installation of the global mode, so the measured envelope is the amplitude of the global mode.

The separation between IA and IC regions in X_A can be assimilated to a virtual permeable boundary of the system. Further downstream we have one point where the base flow profile becomes linearly stable because of the relaxation of the shear profile. Increasing part of the envelope (between bluff body and X_{max}) defines a wave propagation front, with a thickness $\sim \varepsilon^{-1/2}$. If this length is smaller than the distance between the bluff body and the X_A coordinate, it is to be expected that X_{max} coincides with the thickness of the wavefront [21]. On the other hand if X_A is smaller than this, we observe a strong inhomogeneity situation ("small box") but without the same scaling laws. This is always the case for very small values of ε as appears in confined system instabilities, where the correlation length is about the size of the box [7].

11.3 Current Developments

This short summary of results concerns the spatial inhomogeneities in the open systems instabilities as Bénard–von Karman instabilities.

Other problems of interest are the spatial amplitude of the nonlinear stationary mode or zero frequency mode, which is of major importance in unstable flows. Indeed, one of the most significant nonlinearities is the quadratic one that generates a zero frequency mode (stationary) of order ε, describing the deformation of the base flow caused by instability [22, 23, 24]. In the case of wake (Bénard - von Karman instability), the spatial distribution or envelope of this zero mode is concentrated on the axis, in the recirculation region behind the bluff body [16]. The effect of this localization is to pump the fluid inside the region of the negative velocity to oppose recirculation. For stationary flow (without vortex emission), between the existence of recirculation loops ($R \sim 5$) and until the appearance of vortex shedding ($R = R_c = 47$), the recirculation length L_{ro} increases with flow velocity U_o [25, 26] according to $L_{ro} \sim U_o$. The existence of

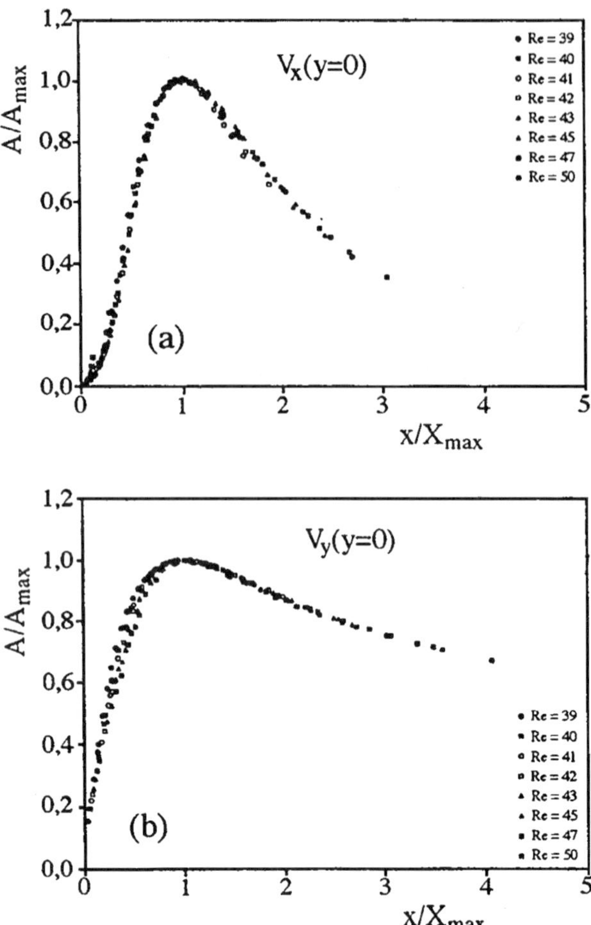

Fig. 11.4. Renormalization curve of the envelope for the (a) V_x and (b) V_y components [16].

the zero-frequency mode decreases locally the flow velocity in this region in a quantity proportional to ε, so that for Reynolds numbers $R > R_c$, the size of the recirculation area L_{rv}, in the presence of the vortex emission should vary as

$$L_{rv} \sim (U_o - Cte.\varepsilon),$$

a phenomenon that we have been able to put in evidence [27]. In addition, this mode modifies the mean drag [28, 29].

Finally, recent investigations on the wakes behind a bluff body have been concerned with developed turbulence situations. Their interest lays in the inhomogeneous character of the turbulence that involves the impact of coherent structures in their statistics. The latter are more significant in near–wake than in far–wake where turbulence is more homogeneous. In this way intermittency exponents [30] measuring deviation from the predictions of Kolmogorov theories in 1941 exhibit a spatial variation with a stronger intermittency in the near–wake tending towards values characteristic of homogeneous turbulence in the far–wake [31, 32].

Acknowledgments

The first part of this work (Rayleigh–Bénard convection) was done by J. E. Wesfreid with Pierre Bergé, Monique Dubois, Christianne Normand, and Yves Pomeau in Saclay. Work on the wake was developed at the ESPCI, with Patrice Jenffer, Barbara Zielinska, Jan Ducek, Eric Gaudin, Bartosz Protas, and Jan Wojciechowski. We thank them for this collaboration.

References

1. P. Bergé, Y. Pomeau, and Ch. Vidal, *Order Within Chaos* (Hermann, Paris, 1984).
2. P. Bergé, Y. Pomeau, and Ch. Vidal, *L'espace Chaotique* (Hermann, Paris, 1998).
3. B. Castaing, G. Gunaratne, F. Heslot, L. Kadanoff, A. Libchaber, S. Thomae, X. Wu, S. Zaleski, and G. Zanetti, Scaling of hard thermal turbulence in Rayleigh–Bénard convection, *J. Fluid. Mech.* **204**, 1 (1989).
4. J.E. Wesfreid, Henri Bénard (1874–1939): Biographie scientifique, *Bull. S.F.P. suppl.* **81**, 39 (1991) and see also Chapter 2 in this book.
5. J.E. Wesfreid, Y. Pomeau, M. Dubois, C. Normand, and P. Bergé, Critical effects in Rayleigh–Bénard convection, *J. Phys. (France)* **39**, 725 (1978).
6. J.E. Wesfreid, P. Bergé, M. Dubois, Induced pre-transitional Rayleigh–Bénard convection, *Phys. Rev A* **19**, 1231 (1979).
7. J.E. Wesfreid, *Etude de la convection de Rayleigh–Bénard au voisinage du seuil de convection*, Thèse de doctorat d'Etat, Université de Paris-Sud, 1981.
8. P. Manneville, *Dissipative Structures and Weak Turbulence* (Academic Press, Boston, 1990).
9. Y. Pomeau and P. Manneville, Stability and fluctuations of a spatially periodic convective flow, *J. Phys. Lett.* **40**, 609 (1979).
10. J.E. Wesfreid and V. Croquette, Forced phase diffusion in Rayleigh–Bénard convection, *Phys. Rev. Lett.* **45**, 634 (1980).
11. J.E. Wesfreid and S. Zaleski(eds), *Cellular Structures in Instabilities*, Lecture Notes in Physics **210**, (Springer, Berlin 1984).
12. C. Mathis, M. Provansal, L. Boyer, The Bénard-von Karman instability: An experimental study near the threshold, *J. Phys. Lett.* **45**, L-483 (1984).
13. M.M. Zdravkovich, *Flow Around Circular Cylinders* (Oxford Science Publications, 1997).
14. C. Mathis, Propriétés des composantes de vitesse tranverse dans l'écoulement de Bénard–von Karman aux faibles nombres de Reynolds, *Thèse de doctorat de l'Univeristé d'Aix-Marseille*, 1983.

15. S. Goujon-Durand, J. E. Wesfreid, and P. Jenffer, Downstream evolution of the Bénard–von Karman instability, *Phys. Rev. E* **50**, 308 (1994).

16. J.E. Wesfreid, S. Goujon-Durand, and B.J.A. Zielinska, Global mode behavior of the streamwise velocity in wakes, *J. Phys. II* (France) **6**, 1343 (1996).

17. B.J.A. Zielinska and J.E. Wesfreid, On the spatial structure of global modes in wake flow, *Phys. Fluids* **7**, 1418 (1995). See also B. Thiria, *Propriétés dynamiques et de stabilité dans les écoulements ouverts forcés*, Thèse de doctorat de l'Université de Paris VI, 2005.

18. X. Yang and A. Zebib, Absolute and convective instability of a cylinder wake, *Phys. Fluids A* **1**, 689 (1989).

19. K. Hanneman and H. Oertel Jr, Numerical simulation of the absolutely and convectively unstable wake, *J. Fluid Mech.* **199**, 55 (1989).

20. P. Huerre and P.A. Monkewitz, Local and global instabilities in spatially developing flows, *Annual Rev. Fluid Mech.* **22**, 473 (1990).

21. A. Couairon and J.M. Chomaz, Fully nonlinear global modes in slowly varying flows, *Phys. Fluids* **11**, 3688 (1999).

22. J. Dusek, P. le Gal, and Ph. Fraunié, A numerical and theoretical study of the first Hopf bifurcation in a cylinder wake, *J. Fluid Mech.* **264**, 59 (1994).

23. V. Pagneux, A. Maurel, Etude numérique d'instabilités en écoulements ouverts confinés, *Comp. Rend. Acad. Sci.* (Paris) **319**, 27 (1994).

24. A. Maurel, V. Pagneux, and J.E.Wesfreid, Mean-flow correction as nonlinear saturation mechanism, *Europhys. Lett.* **32**, 217 (1995).

25. M. Nishioka and H. Sato, Mechanism of determination of the shedding frequency of vortices behind a cylinder at low Reynolds numbers, *J. Fluid Mech.* **89**, 49 (1978).

26. M. Coutanceau and R. Bouard, Experimental determination of the main features of the viscous flow in the wake of a circular cylinder in uniform translation. I - Steady flow, *J. Fluid Mech.* **79**, 231 (1977).

27. B.J.A. Zielinska, S. Goujon-Durand, J. Dusek, and J.E. Wesfreid, Strongly nonlinear effect in unstable wakes, *Phys. Rev. Lett.* **79**, 3893 (1997).

28. B. Protas and J.E. Wesfreid, Drag force in the open-loop control of the cylinder wake in the laminar regime, *Phys. Fluids* **14**, 810 (2002).

29. B. Protas and J.E. Wesfreid, On the relation between the global modes and the spectra of drag and lift in periodic wake flows, *C.R. Mécanique* **331**, 49 (2003).

30. U. Frisch, *Turbulence: The Legacy of A.N. Kolmogorov* (Cambridge University Press, New York 1995).

31. E. Gaudin, B. Protas, S. Goujon-Durand, J. Wojciechowski, and J.E. Wesfreid, Spatial properties of velocity structure functions in turbulent wake flows, *Phys. Rev. E* **57**, R9 (1998).

32. B. Protas, S. Goujon-Durand, and J.E. Wesfreid, Scaling properties of two-dimensional turbulence in wakes behind bluff bodies, *Phys. Rev. E* **55**, 4165 (1997).

Part V

Extension of Bénard's Work

12 Patterns and Chaotic Dynamics in Faraday Surface Waves

Jerry P. Gollub

[1] Physics Department, Haverford College, Haverford, PA 19104, USA
[2] Department of Physics and Astronomy, University of Pennsylvania, Philadelphia, PA 19104, USA

A brief review is given of ordered and disordered patterns formed on the surface of a fluid layer subjected to vertical oscillation. We point out connections to cellular Bénard patterns, and discuss the extent of our understanding of these nonlinear states.

12.1 Introduction

The cellular structures studied by Bénard in thermal convection have parallels in many other systems. One that has been studied prominently is a large fluid layer with a free surface, subjected to vertical oscillation of the container at a fixed frequency. When the excitation acceleration exceeds a frequency-dependent threshold, patterns form that can include stripes, squares, hexagons, and even quasi-crystalline patterns, depending on the frequency, viscosity, and the driving waveform.

Michael Faraday [1] first studied these surface wave patterns in his 1831 paper "On a Peculiar Class of Acoustical Figures, and on the Forms of Fluids Vibrating on Elastic Surfaces". This work was quite varied, and included observations of granular material, "white of egg", alcohol, and milk among others. He noted that milk was advantageous for determining the form of the waves because of its light scattering properties. The experiments were surprisingly ambitious: a large version used "a board eighteen feet long" and a layer of water "twenty-eight inches by twenty inches in extent". He recognized that sloping the bottom near the boundaries of the container could prevent reflections. In other words, the experiments were remarkable for their time.

Faraday waves are attractive for studying cellular patterns for several reasons [2]. The basic time scale for pattern evolution is much shorter than is the case for thermal convection, which allows multiple parameters to be studied in a reasonable time. Furthermore, the wavenumber of the pattern can be controlled by the imposed forcing, thus allowing the differences between capillary and gravity waves to be studied. All of the basic pattern symmetries of the plane can be realized. However, the theoretical difficulty of the problem is much harder than is the case for convection because of the free surface. Also, numerical computations are particularly demanding because the primary pattern is oscillatory, a fact that imposes demanding requirements on time resolution. Another difficulty is the fact that lateral boundaries are important if the viscosity is low, so it is difficult to reach the limit of an effectively infinite layer.

Since these early experiments, Faraday waves have been extensively explored, and the subject is briefly summarized here. The reviews of Cross and Hohenberg [3] and Miles and Henderson [4] provide a good starting point. The reader is also referred to more extensive descriptions of regular waves [2] and quasi-crystalline waves [5] that are available elsewhere. We begin with some history, proceed to a discussion of regular wave patterns, and consider their description by amplitude equations, including the effects of wave interactions on the dynamics. We then consider more complex patterns arising in multifrequency forcing, such as quasi-crystals and superlattices. We close with a brief discussion of spatio-temporal chaos of parametric waves. The discussion given at the Workshop is roughly parallel to an invited lecture given at the Centennial Meeting of the American Physical Society. An audiotape and the visuals of this presentation are available on the Web [6].

Faraday's initial observations were unexplained. In 1883, Lord Rayleigh suggested that the waves result from parametric resonance, as in a damped harmonic oscillator of the following form:

$$\frac{d^2x}{dt^2} + 2\mu\frac{dx}{dt} + \omega_o^2(t)x = 0, \tag{12.1}$$

where the natural angular frequency ω_o is modulated at a frequency f, and the resulting oscillation occurs at a frequency $f/2$.

In 1954, Benjamin and Ursell [7] showed from the inviscid Euler equations that each normal mode of the container (for small amplitudes) acts as a harmonic oscillator with time-dependent frequency determined by the dispersion relation for capillary–gravity waves:

$$\omega_o^2(t) = [g(t)k + \frac{\sigma k^3}{\rho}]\tanh kh \tag{12.2}$$

Here, $g(t)$ represents the effective oscillation of the gravitational field resulting from the vibration, h is the depth of the fluid layer, and k is the wavenumber.

Damping is very difficult to include properly. The stability problem with damping was solved in 1994 by Kumar and Tuckerman [8] using Floquet analysis. The finite damping causes the instability to occur at finite driving acceleration, and the predictions are in good agreement with experiment [9].

12.2 Primary Patterns

Phase diagrams showing the patterns observed as a function of acceleration and frequency have been published for sinusoidal forcing by Kudrolli and Gollub [2]. The general behavior is as follows. Stripes are found at the highest viscosities (above 1 cm^2/s or so) over a wide range of frequencies. These often show deformations similar to those found for thermal convection, due to boundary effects that penetrate far into the interior, as shown in Figure 12.1. At lower viscosities, squares are found for frequencies above 40 Hz or so, and hexagons dominate for lower frequencies. The hexagons are quite regular, and have a special feature not found for thermal convection. Because the waves are subharmonic relative to the forcing frequency, there are two possible phases of the patterns relative to the forcing, and hence phase defects can form, with part of the pattern having one phase, whereas the remainder has the other pattern, as shown in Figure 12.2.

Fig. 12.1. Primary striped wave pattern formed at high viscosity $\nu = 1.0\,cm^2 s^{-1}$ and frequency $f = 45Hz$ at large aspect ratio (from [2]).

At very low viscosities and a narrow range of moderate frequencies for which both capillarity and gravity are significant, quasi-crystalline patterns can be found, as has been demonstrated by Binks and van de Water [10].

Patterns containing domains of different symmetry can coexist. For example, squares, stripes, and hexagons are sometimes found together, as shown in Figure

12.3. This may be a consequence of the inhomogeneity induced by finite system size.

Fig. 12.2. Hexagonal wave pattern showing a phase defect ($\nu = 0.2 \; cm^2 \; s^{-1}, f = 20$ Hz). The lighter and darker regions have different phases of oscillation relative to the external modulation (from [2]).

Nonlinear evolution equations for standing wave patterns with amplitude B_1 interacting with other waves of amplitude B_m were derived by Chen and Viñals [11] for periodic forcing:

$$\frac{dB_1}{dt} = \alpha B_1 - g_o B_1^3 - \sum_{m \neq 1} g(\theta_{m1}) B_m^2 B_1, \tag{12.3}$$

where α and g_o are constants, and the quantities $g\theta_{m1}$ are interaction coefficients that depend on the angles between two waves. The amplitude equation is variational, minimizing the value of a certain functional F:

$$F = -\frac{1}{2}\,\alpha \sum_m B_m^2 + \frac{1}{4} \sum_{m \neq n} g(\theta_{mn}) B_m^2 B_n^2. \qquad (12.4)$$

There is no quadratic term in the amplitude equationas a consequence of symmetry: The fluid equations are invariant under time translation by one forcing period, but this reverses the sign of B, so both B and $-B$ must be solutions. Therefore there can be no quadratic term. In thermal convection, hexagons are explained by quadratic terms in the amplitude equation, so that a completely different explanation is required here. Wave patterns near onset are determined by the angular dependence of the nonlinear coupling coefficient $g(\theta)$, and this gives rise to transitions as viscosity and driving frequency are varied.

Fig. 12.3. Weakly time-dependent coexistence of hexagons, squares, and stripes in surface waves ($\nu = 0.5 cm^2 s^- 1$; $f = 32.5 Hz$) (from [2]).

The key physical effect determining the structure of $g(\theta)$ (and hence the preferred pattern) is the presence of three-wave resonances between different modes, which are affected by the shape of the wave dispersion relation. The frequencies and wavenumbers of the three interacting waves must satisfy conservation laws

$$\omega_1 + \omega_2 = \omega_3 \ and \ \boldsymbol{k}_1 + \boldsymbol{k}_2 = \boldsymbol{k}_3 \tag{12.5}$$

as well as the dispersion relation of Equation (12.2). For capillary waves, the preferred angle for nonlinear interactions is $\theta = 75°$. Two waves at frequency ω near this angle lose energy to a wave at 2ω and are disfavored. On the other hand, waves near $90°$ do not interact much and are hence favored, giving rise to squares at low viscosity. As the viscosity is increased, a transition to stripes is predicted and observed.

At lower frequency, where gravity accounts for most of the restoring force, the resonant angle approaches zero, so waves at other angles do not interact much, and it therefore pays to have more of them, according to the first term in Equation (12.4). In this way, hexagons (a superposition of three real waves at $60°$ relative angles) are favored over a fairly wide range of conditions. For a much narrower range of conditions, quasi-crystalline patterns, with more than three superposed real plane waves, are favored, consistent with the experimental observations mentioned above.

12.3 Two-Frequency Forcing

Forcing simultaneously at two incommensurate frequencies (e.g., 4ω, 5ω) allows quadratic terms to appear in the amplitude equation, and this can change the favored symmetries of ordered patterns. In particular, hexagons are favored even for pure capillary waves, which is not possible via the angular dependence of the nonlinear coefficient in the amplitude equation.

Nonlinear wave interactions can lock various standing wave patterns together to produce novel lattices. For example, the interaction of two hexagonal lattices at a $30°$ relative orientation (i.e., six standing waves at $30°$) gives rise to a quasi-crystalline pattern of twelvefold orientational symmetry, as shown in Figure 12.4. This pattern was discovered by Edwards and Fauve [12], and was later studied in [5]. The twelvefold symmetry can be seen by sighting along the figure at a glancing angle. However, there is in principle no precise translational symmetry; instead, the pattern comes arbitrarily close to repeating given a large enough translation. Of course, in a finite system, quasi-periodicity is something of a fiction. The transition from hexagons to the quasi-crystalline state is found to be discontinuous.

Other patterns with distinct types of symmetry are also found for two-frequency forcing. These are called superlattice patterns [5] because they involve several apparent length scales. One of them, shown in Figure 12.5 and called *superlattice-1*, is a spatially periodic pattern that can be viewed as consisting of two hexagons oriented at a relative angle of $2\ sin^{-1}(1/2\sqrt{7}) \cong 22°$. The spatial phases of the pattern are also important. A point of sixfold symmetry of the first hexagonal lattice coincides with a point of triangular symmetry of the second. Although this first superlattice pattern was found for frequency ratio 6:7,

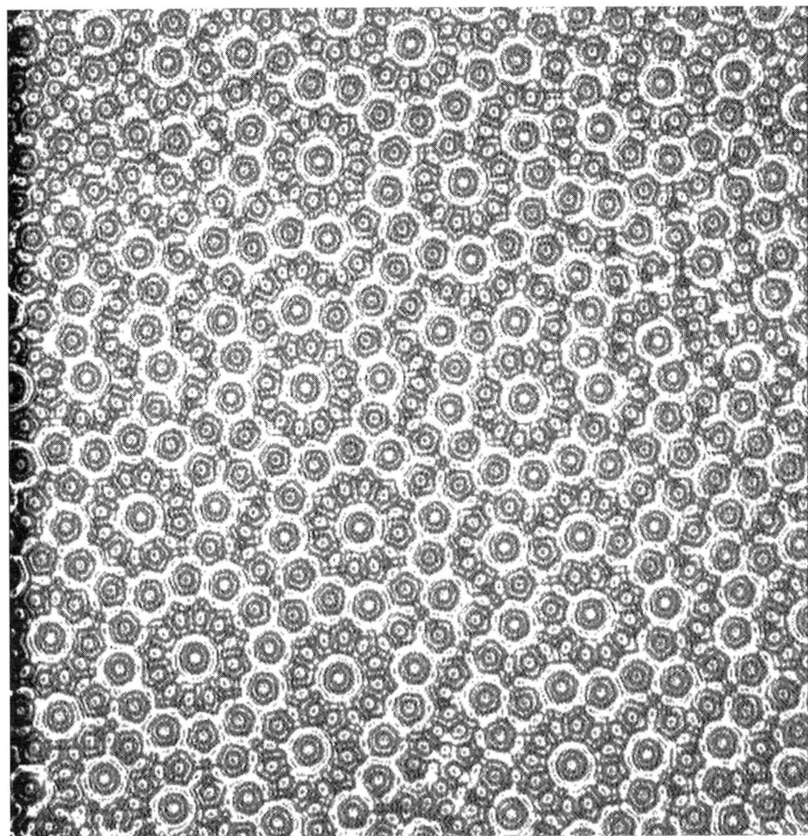

Fig. 12.4. A quasi-crystalline wave pattern with twelvefold orientational symmetry, produced by forcing a layer of silicone oil simultaneously at two frequencies, using a method invented by Edwards and Fauve (from [12]).

a different one is found for ratio 4:5. The time average of this "superlattice-2" pattern (Figure 12.6) can be represented as a combination of two hexagonal lattices differing in wavenumber by a factor $\sqrt{3}$. However, the sixfold symmetry visible in the time average pattern is broken at any particular instant. Superlattice patterns have also been investigated by Arbel and Fineberg [13], who discovered further novel variations. Silber and Proctor [14] and Porter and Silber [15] have shown that considerable insight into these superlattice patterns can be obtained by using symmetry arguments. Lifshitz and Petrich [16] also modeled superlattice patterns.

These examples by no means exhaust the possibilities. Figure 12.7 shows a final example, a snapshot of a hexagon/stripe mixed state that is strikingly regular. In this example, the amplitudes of the three coupled standing waves are unequal, with one apparently predominating.

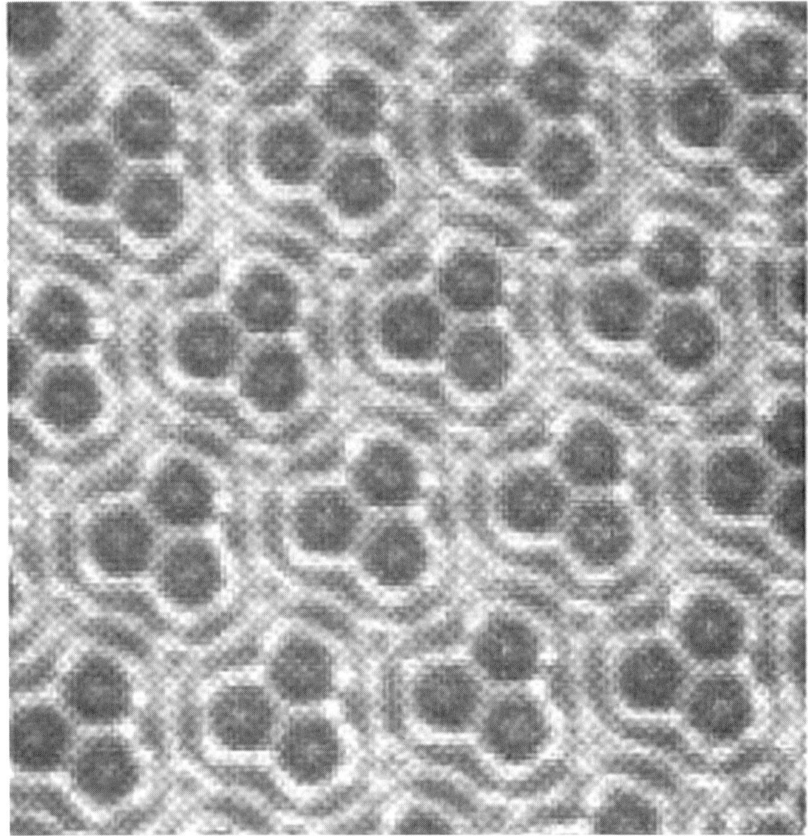

Fig. 12.5. Superlattice-I pattern obtained for two-frequency forcing of Faraday waves with frequency ratio 6:7(magnified) (from [5]).

In summary, the regular patterns of parametrically forced surface waves arise from nonlinear interactions between standing wave components characterized by different wavevectors. An amplitude equation approach, in conjunction with symmetry considerations, gives a quantitative explanation of the regular patterns for single frequency forcing. Distinct pattern symmetries occur as parameters are varied, modifying the allowed angles for three-wave resonances. The more exotic patterns (e.g., superlattices involving coupled hexagonal patterns formed by multiple frequency forcing) appear to be understood qualitatively.

12.4 Spatio–Temporal Chaos and Other Complex Phenomena

Now we turn to the topic of spatio–temporal chaos (STC) of parametrically forced waves. This phenomenon has been studied for a long time, starting with

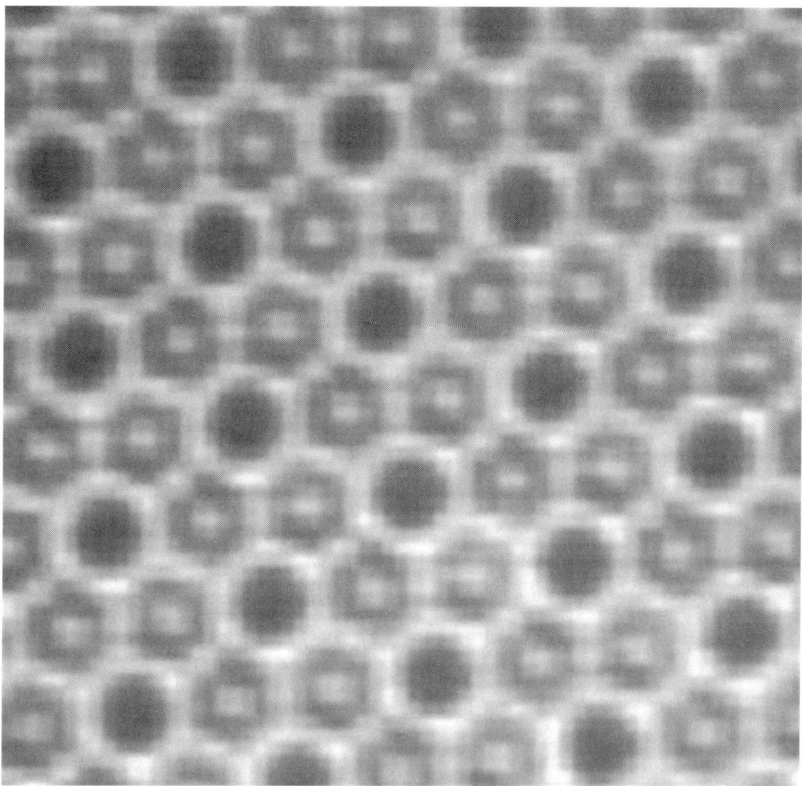

Fig. 12.6. Superlattice-II pattern produced for two-frequency forcing of Faraday waves with frequency ratio 4:5 (magnified) (from [5]).

Ezersky et al. [17], where an amplitude modulation instability of square patterns was shown to lead to STC.

Given that there are a host of primary patterns, and the fact that the stability of the primary pattern must affect the transition to STC, it is evident that there must be a host of mechanisms for producing STC. Three particular cases are: (a) amplitude modulations of stripes, described theoretically by Milner [18]; (b) a mixed state in which order and spatiotemporal chaos coexist in different regions of the cell [19]; and (c) a continuous melting transition of hexagons with increasing excitation [2]. The last of these processes was hypothesized to involve a breakdown of the coupling of the standing wave modes to each other, but this has not been proven.

At high excitation and low viscosity, a state of "wave turbulence" occurs with a very broad wavenumber spectrum [20] that appears to decay somewhat faster than k^{-4} at high wavenumbers. The observations were compared to a hydrodynamic cascade model whose predictions were fairly similar to the observations.

Another related research direction involves the study of waves in nontraditional fluids, such as granular materials. Many of the patterns seen in simple fluids occur in granular material when a layer is subjected to vertical vibration, as shown by Melo, Umbanhowar, and Swinney [21]. In addition, there are striking localized excitations that seem to arise from a featureless background [22]. Similar localized excitations can be formed in colloidal suspensions [23].

The subject of parametrically forced surface waves is extraordinarily rich. In this brief review, we have seen that surface wave patterns have interesting parallels with Bénard convection. In both cases, a variety of cellular patterns are formed that can be explained near onset using coupled amplitude equations. In more complex nonlinear situations, novel ordered states such as quasi-crystalline patterns and superlattices are formed. A variety of mechanisms leading to spatio–temporal chaos and wave turbulence also occur. Some of these complex phenomena are partially understood, whereas others place difficult demands on the theory at its present state of development.

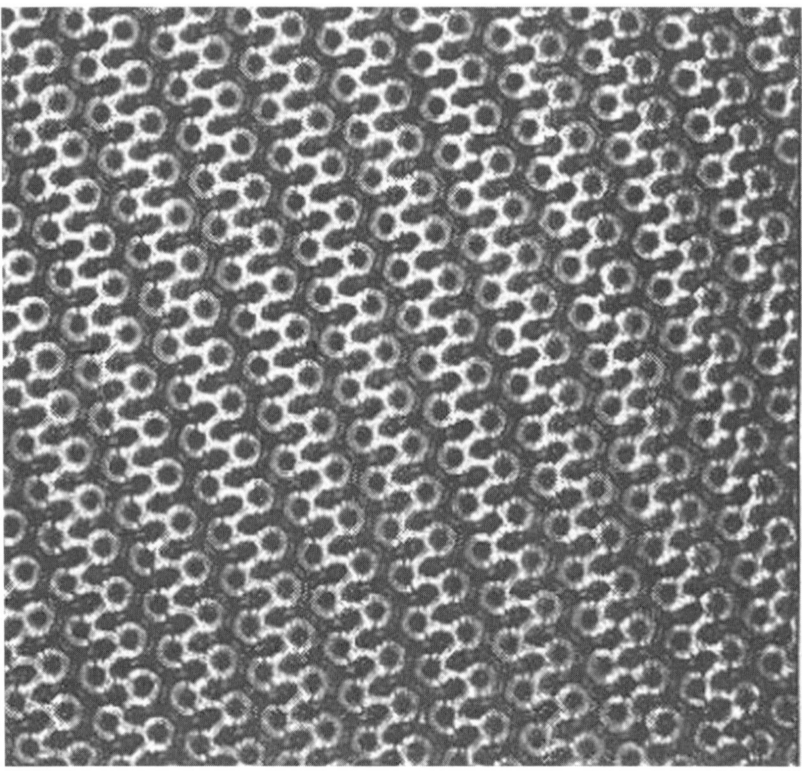

Fig. 12.7. Hexagon/stripe mixed state where the amplitude of one of the three standing wave components of the hexagons is enhanced.

Acknowledgments

We appreciate financial support from the National Science Foundation Division of Materials Research under Grant DMR-0072203. Arshad Kudrolli, Stuart Edwards, Benoit Pier, Bruce Gluckman, and Craig Arnold contributed to the work done at Haverford which is described here.

References

1. M. Faraday, On a Peculiar Class of Acoustical Figures, and on the Forms of Fluids Vibrating on Elastic Surfaces, *Philos. Trans. R. Soc. London* **121**, 299 (1831).
2. A. Kudrolli and J.P. Gollub, Patterns and spatiotemporal chaos in parametrically forced surface waves: a systematic survey at large aspect ratio, *Physica* D **97**, 133 (1996).
3. M.C. Cross and P.C. Hohenberg, Pattern formation outside of equilibrium, *Rev. Mod. Phys.* **65**, 851 (1993).
4. J. Miles and D. Henderson, Parametrically forced surface waves, *Annu. Rev. Fluid Mech.* **22**, 143 (1990).
5. A. Kudrolli, B. Pier, and J.P. Gollub, Superlattice Patterns in Surface Waves, *Physica* D **123**, 99 (1998).
6. For an audiotape and the transparencies of this presentation, please see http://www.apscenttalks.org/pres_masterpage.cfm?nameID=68.
7. T.B. Benjamin and F. Ursell, The stability of the plane free surface of a liquid in vertical periodic motion, *Proc. R. Soc. London* A **225**, 505 (1954).
8. K. Kumar and L.S. Tuckerman, Parametric instability of the interface between two fluids, *J. Fluid Mech.* **279**, 49 (1994).
9. J. Bechhoeffer, V. Ego, S. Manneville, and B. Johnson, An experimental study of the onset of parametrically pumped surface waves in viscous fluids, *J. Fluid Mech.* **288**, 325 (1995).
10. D. Binks and W. van de Water, Nonlinear pattern formation of Fraday waves, *Phys. Rev. Lett.* **78**, 4043 (1997).
11. P. Chen and J. Viñals, Pattern selection in Faraday waves, *Phys. Rev. Lett.* **79**, 2670(1997).
12. W.S. Edwards and S. Fauve, Patterns and quasi-patterns in the Faraday experiment, *J. Fluid Mech.* **278**, 123 (1994).
13. H. Arbell and J. Fineberg, Two-mode rhomboidal states in driven surface waves, *Phys. Rev. Lett.* **84**, 654 (2000).
14. M. Silber and M.R.E. Proctor, Nonlinear competition between small and large hexagonal patterns, *Phys. Rev. Lett.* **81**, 2450 (1998).
15. J. Porter and M. Silber, Broken symmetries and pattern formation in two-frequency forced Farady waves, *Phys. Rev. Lett.* **89**, 084501 (2002).
16. R. Lifshitz and D.M. Petrich, Theoretical model for Faraday waves with mulitple-frequency forcing, *Phys. Rev. Lett.* **79**, 1261 (1997).
17. A.B. Ezersky, M.I. Rabinovich, V.P. Reutov, and I.M. Starobinets, Spatiotemporal chaos at parametric excitation of capillary ripples, *Zh. Eksp. Teor. Fiz.* **91**, 2070 (1986).
18. S.T. Milner, Square patterns and secondary instabilities in driven capillary waves, *J. Fluid Mech.* **225**, 81 (1991).

19. A. Kudrolli and J.P. Gollub, Localized Spatiotemporal Chaos in Surface Waves, *Phys. Rev. E* **64**, R1052 (1996).
20. W.B. Wright, R. Budakian, and S.J. Putterman, Diffusing light photography of fully developed isotropic ripple turbulence, *Phys. Rev. Lett.* **76**, 4528 (1996).
21. F. Melo, P.B. Umbanhowar, and H.L. Swinney, Transition to parametric wave patterns in a vertically oscillated granular layer, *Phys. Rev. Lett.* **72**, 172 (1994).
22. P.B. Umbanhowar, F. Melo, and H. Swinney, Localized excitaions in a vertically vibrated granular layer, *Nature* (London) **382**, 793 (1996).
23. O. Lioubashevski, H. Arbell, and J. Fineberg, Dissipative solitary states in driven surface waves, *Phys. Rev. Lett.* **76**, 3959 (1996).

13 The Taylor–Couette Flow: The Hydrodynamic Twin of Rayleigh–Bénard Convection

Arnaud Prigent[1], Bérengère Dubrulle[2], Olivier Dauchot[2], and Innocent Mutabazi[1]

[1] Laboratoire de Mécanique, Physique et Géosciences,
  Université du Havre, 25 Rue Philippe Lebon, F–76058 Le Havre Cedex, France
[2] Groupe Instabilités et Turbulence,
  DSM/SPEC, CEA Saclay, F–91191 Gif sur Yvette Cedex, France

There is a strong analogy between Rayleigh–Bénard convection and the Taylor–Couette system. This analogy is well known when dealing with the primary instability, and is based on the existence of an unstable stratification in both systems. We show that the analogy can be extended beyond the primary instability modes, to the weakly non-linear regime and even further to the fully turbulent one.

13.1 Introduction

Early studies of the Taylor–Couette flow, the flow between two differentially rotating coaxial cylinders [1, 2], are contemporaries of Bénard's works on convection. Very soon, a strong analogy between both sytems was observed [3]. In both cases the flow developing at the onset of instability is made of vortices called, respectively, *Taylor vortices* and *Bénard rolls* (Figure 13.1).

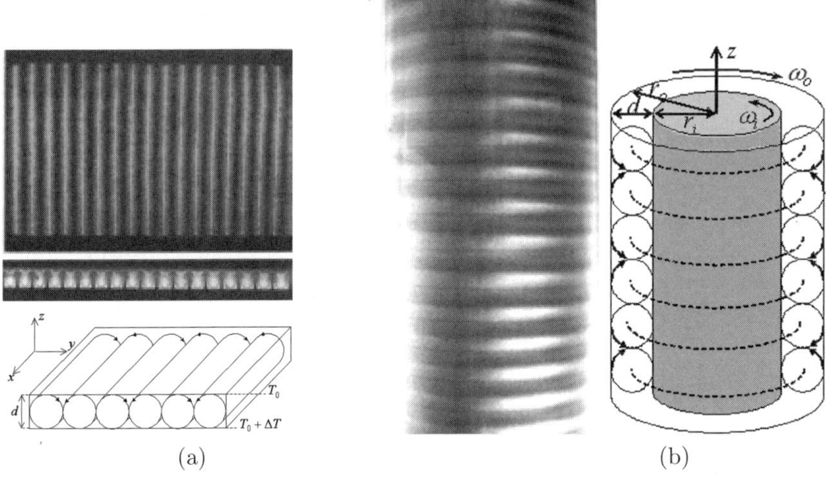

Fig. 13.1. Pictures and schematic views of (a) the Rayleigh–Bénard rolls (courtesy of F. Daviaud from SPEC-CEA, Saclay) and (b) the Taylor vortices.

The first studies of the Taylor–Couette flow consisted of viscosity measurements performed by Couette [1] and by Mallock [2]. In order to avoid the appearance of instablities, Couette rotated the outer cylinder and fixed the inner one. The instability of rotating flows was then considered by Rayleigh in a theoretical paper [3] published the same year as his work on convection [4]. A few years later, Taylor [5], conducting both experimental and theoretical works, re-examined the whole Couette flow problem. Using an apparatus where both cylinders could rotate independently, he predicted theoretically and verified experimentally the instability thresholds for both co- and counterrotating regimes. He showed that the flow, now called the *Taylor vortex flow*, developing at onset of instability is axisymmetric, periodic in the axial direction, and quantitatively similar to Bénard rolls of convection. This strong similarity is the signature of the analogy between the linear stability properties of rotating flows and the stability of stratified fluid first reported by Rayleigh and subsequently analyzed by many authors [6, 7, 8, 9, 10].

In this chapter, we review the analogy between the two systems and extend it beyond the primary instability modes. In the second section, we present the basic states of both systems and the physical mechanisms of their destabilization, which lead us to the detailed formulation of the analogy. The third section addresses the weakly nonlinear and the secondary instability modes. Finally, in the last section, we discuss the implications of the extension of the analogy to the turbulent regime. More specifically, we show how it allows us to predict precise scaling laws for the turbulent transport properties in the Taylor–Couette flow, on the basis of precise measurements conducted in the turbulent Rayleigh–Bénard convection.

13.2 Basic Flows and Instability Mechanisms

The Rayleigh–Bénard convection is obtained when a fluid layer of thickness d is confined between two horizontal plates maintained at different temperatures, the bottom plate being set to a higher temperature $T_0 + \Delta T$ and the top one kept at the temperature T_0. In the presence of the gravity field $\boldsymbol{g}$, the temperature gradient, due to the difference in temperature between the two plates, induces a vertical density stratification. The flow is controlled by the Rayleigh number $Ra = \alpha \Delta T g d^3 / \nu \kappa$, where α is the thermal expansion coefficient, ν is the kinematic viscosity, and κ is the thermal diffusivity. For small Ra the fluid is at rest, the temperature transport is purely diffusive, and the density stratification remains stable (conduction regime). There exists a critical value Ra_c above which instability occurs and Rayleigh–Bénard rolls develop (Figure 13.1a).

The Taylor–Couette flow is realized with a fluid enclosed in the gap between two independently rotating coaxial cylinders of common length L. If $r_{i,o}$ and $\omega_{i,o}$ are the radius and the angular velocity of the inner and outer cylinders, then choosing $d = r_o - r_i$ as unit length and d^2/ν as unit time, the natural control parameters are the radius ratio $\eta = r_i/r_o$, the aspect-ratio $\Gamma = L/d$, and the Reynolds numbers defined for the inner and outer cylinders, respectively:

$R_{i,o} = r_{i,o}\omega_{i,o}d/\nu$. In theoretical studies, it is sometimes convenient to replace the Reynolds numbers by another set of parameters (μ, Ta) where $\mu = \omega_o/\omega_i$ is the rotation ratio and Ta is the *Taylor number* that takes into account the inertia, dissipation and curvature effects (i.e., $Ta \sim Ri^2(d/r_i)$). For a given geometry, η and Γ are constant and the primary instability threshold is given by the curve $Ta_c(\mu)$ in the parameter space (μ, Ta). For $Ta < Ta_c$, the basic flow, called the *circular Couette flow*, is purely azimuthal and depends only on the radial coordinate r. The velocity and the pressure are given by

$$v_\theta = V(r) = Ar + \frac{B}{r}$$

and

$$P(r) = P_0 + \rho \int \frac{V^2(r)}{r} dr,$$

where the constants A and B are determined by the no-slip boundary conditions $v_\theta(r = r_{i,o}) = r_{i,o}\omega_{i,o}$:

$$A = \frac{\omega_i \left(\mu - \eta^2\right)}{1 - \eta^2} \quad \text{and} \quad B = \frac{\omega_i r_i^2 \left(1 - \mu\right)}{1 - \eta^2}, \tag{13.1}$$

where $\mu = \omega_o/\omega_i$. For $Ta > Ta_c$, instability occurs and the Taylor vortices appear (Figure 13.1b). One can already note at this stage that the circular Couette flow is independent of the fluid viscosity in the same way the resting base state in Rayleigh–Bénard convection is independent of the thermal diffusivity. Also, as first shown by Jeffreys in 1928 [6], the analogy between both flows goes far beyond the similarity evidenced in figure 13.1. In order to make this analogy explicit, we now turn to the linear stability analysis of the basic states, focusing on the Taylor–Couette case and referring to the chapter 3 by Manneville in the present volume for Rayleigh–Bénard convection. Let us consider an infinitesimal axisymmetric perturbation of the form $\boldsymbol{v}'(r, z) = (v_r', v_\theta', v_z')$ and $p'(r, z)$. The linearized Navier–Stokes and continuity equations read:

$$\begin{aligned} \boldsymbol{\nabla}.\boldsymbol{v}' &= 0 \\ \left(\tfrac{\partial}{\partial t} + M\right) \boldsymbol{v}' &= -\boldsymbol{\nabla}p' + \nu\Delta\boldsymbol{v}', \end{aligned} \tag{13.2}$$

where

$$\mathbf{M} = \begin{pmatrix} 0 & -\frac{2V}{r} & 0 \\ \frac{dV}{dr} + \frac{V}{r} & 0 & 0 \\ 0 & 0 & 0 \end{pmatrix} \tag{13.3}$$

is the inertial operator for axisymmetric inviscid perturbations. As a result, one obtains that the flow is linearly stable when the *Rayleigh discriminant*

$$\phi(r) = \frac{2V}{r} \left(\frac{dV}{dr} + \frac{V}{r}\right)$$

is positive [11, 12]. The above equations are indeed similar in structure to the linearized Boussinesq equations for Rayleigh–Bénard convection [13]. When expanding the perturbation field into normal modes, in the small gap approximation, these equations reduce to a single equation for the marginal mode

$v(x) \exp(ikz)$:

$$\left(\frac{d^2}{dx^2} - q^2\right)^3 v(x) = -\left[1 - (1-\mu)x\right] Ta\, q^2 v(x), \qquad (13.4)$$

where $x = (r - r_i)/d$, $q = kd$ and Ta is the Taylor number defined by

$$Ta = \frac{-4A\omega_i d^4}{\nu^2} = 4\frac{1-\eta}{1+\eta} R_i^2 \left(1 - \frac{\mu}{\eta^2}\right). \qquad (13.5)$$

Apart from the radial dependence of the coefficient of $v(x)$ in the r.h.s. term, Equation (13.4) is formally identical to the equation for the temperature perturbation Θ in Rayleigh–Bénard convection as soon as one replaces the z-derivative by the x-derivative and the Rayleigh number by the Taylor number Ta [7]:

$$\left(\frac{d^2}{dz^2} - q^2\right)^3 \Theta(z) = -Ra\, q^2 \Theta(z) \qquad (13.6)$$

In the limit $\mu \to 1$, the Taylor number for which instability occurs is $Ta_c = 1708$ and the corresponding wave number is $q_c = 3.12$ [7]. These values are precisely those obtained in Rayleigh–Bénard convection with rigid–rigid boundary conditions. Outside this limit, the dependence of the instability threshold on the radial coordinate calls for a physical prescription to determine the position at which the instability first takes place. Studying the mechanism of the instability will now provide us with a physical grounding of the analogy and thereby with the desired prescription.

Let us consider a spherical fluid particle with radius R rotating near the inner cylinder in the Couette flow. We introduce a small perturbation in the form of a radial shift of this blob with velocity v (Figure 13.2a). This is the analogue of a temperature perturbation in the form of a vertical shift in the fluid at rest between the two differentially heated plates in the Rayleigh–Bénard case (Figure 13.2b). If this blob experiences forces that tend to reinforce its movement, the flow is unstable. In contrast, if forces tend to bring it back to its initial position, the flow is stable. We first consider the inviscid case. In the Taylor–Couette system, the centrifugal force per unit mass V^2/r acting in the radial direction on a fluid element at position r is balanced by the centripetal radial pressure gradient at this position. Now, consider an outward displacement of the blob, initially at r, towards $r' = r + \delta r$. Owing to axisymmetry, the angular momentum per unit mass $\mathcal{L} = rV$ is a conserved quantity. At its new position, the particle is submitted to a centrifugal force per unit mass $f = \mathcal{L}^2/r'^3$. At this new position, the centripetal pressure gradient is given by the local equilibrium with the centrifugal force $f' = \mathcal{L}'^2/r'^3$. If $\delta f = f' - f > 0$, the pressure gradient pushes back the element to its initial position. On the contrary, if $\delta f < 0$, the fluid continues to move outward until it reaches the outer cylinder where it is forced to come back. Accordingly, the criterion for stability is given by

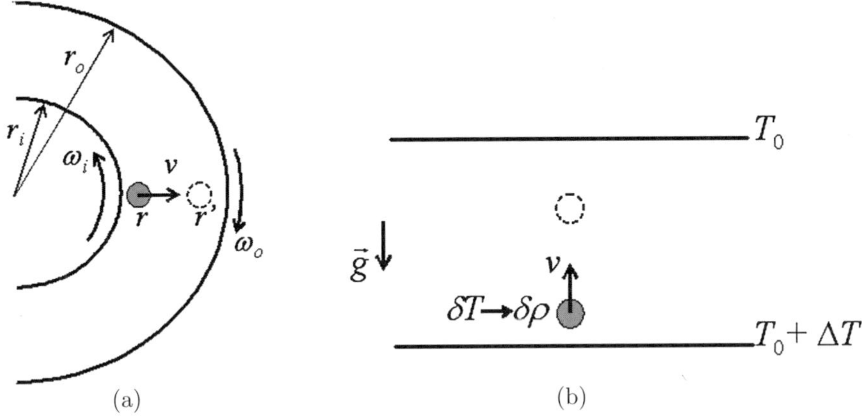

Fig. 13.2. (a) Displaced spherical fluid particle in the laminar circular Couette flow: v stands for the radial velocity v_r of the displaced particle; and (b) in the fluid at rest in a Rayleigh–Bénard cell: v is the vertical velocity of the displaced particle.

$$\frac{\delta f}{\delta r} = \frac{1}{r^3}\frac{d\mathcal{L}^2(r)}{dr} = \frac{2V}{r}\left(\frac{dV}{dr} + \frac{V}{r}\right) = \phi(r) > 0. \tag{13.7}$$

Introducing now the effect of viscosity allows us to formulate this criterion with the Taylor number as defined in equation (13.5). Indeed, the same mechanism applies except that instability may occur only if the destabilizing force $-\delta f$ overpasses the viscous damping per unit mass $f_{\text{visc}} \sim \rho\nu R v_r/\rho R^3$ on a time scale smaller than the viscous time $\tau_\nu \sim \frac{R^2}{\nu}$ which limits the displacement of the blob to $\delta r \lesssim \tau_\nu v_r$. As a result, the criterion for stability becomes

$$\delta f = \phi(r)\delta r \gtrsim -\frac{\nu v_r}{R^2}; \qquad \text{that is,} \qquad \phi(r) \gtrsim -\frac{\nu^2}{R^4}. \tag{13.8}$$

For the azimuthal profile $V(r) = Ar + B/r$, the above stability criterion becomes $4A\Omega(r) \gtrsim -\nu^2/R^4$, which in the most unstable situation ($r = r_i$), and for a blob size $R \sim d$, is equivalent to

$$Ta < \text{cte}, \tag{13.9}$$

where Ta is the Taylor number defined above. The constant on the r.h.s. of the equation depends on geometric factors. It can be adjusted in order to fit the critical value observed in experiments or given by the linear stability analysis.

The mechanism we have described is based on angular momentum stratification in the same way as the mechanism of Rayleigh–Bénard convection is based on vertical density stratification. The driving force of Rayleigh–Bénard convection, the Archimedian buoyancy, is replaced by the net balance of the centrifugal force. The temperature gradient is replaced by the velocity gradient and thermal diffusion is replaced by momentum diffusion. The Taylor number is therefore analogous to the Rayleigh number, the ratio of the Archimedian force over the viscous dissipation force. Note that viscosity acts twice in the Taylor–Couette flow (so the dependence to its square value in the Taylor number): it attenuates velocity gradients as the thermal diffusion attenuates temperature gradients and it dissipates the kinetic energy by friction. The characteristic time associated with the destabilizing centrifugal force is $\tau_c = (1/\omega_i)\sqrt{d/r_i}$; and the viscous dissipation time is $\tau_\nu = d^2/\nu$, therefore the Taylor number can be seen as $Ta \sim (\tau_\nu/\tau_c)^2$ in the same way the Rayleigh number is defined as $Ra = (\tau_\nu/\tau_A) \times (\tau_\kappa/\tau_A)$ where τ_A is the characteristic time of the Archimedian force and $\tau_\kappa = d^2/\kappa$ is the thermal diffusion time. Table 13.1 summarizes the other terms of the analogy.

Table 13.1. Analogy Between Rayleigh–Bénard Convection and Taylor–Couette Flow.

	R.–B. convection	T.–C. system
Coordinates	$z,\ (x,\ y)$	$r,\ (\theta,\ z)$
Viscous dissipation	$\mu v d$	$\mu v d$
Stabilizing characteristic time	d^2/κ	d^2/ν
Driving instability force	Archimedien buoyancy $\rho_0 \alpha g \dfrac{d^5}{\kappa}\dfrac{\Delta T}{d}v$	Centrifugal force $4A\Omega\rho d^3 \dfrac{d^2}{\nu}v$
Destabilizing characteristic time	$\tau_A = (\frac{d^2}{g\alpha\delta T})^{1/2}$	$\tau_c = \frac{1}{\omega}(\frac{d}{r_i})^{1/2}$
Control parameter	$Ra = \alpha\Delta T g d^3/\nu\kappa$	$Ta = -4A\omega_i d^4/\nu^2$
Instability threshold value	$Ra_c = 1708$	$Ta_c = 1708$
Critical wave number	$q_c = 3.12$	$q_c = 3.12$
Perturbation characteristic time τ_0	0.063 $(Pr = 490)$	0.041
Perturbation characteristic length ξ_0	0.385	0.260 $(\eta = 0.883, \Gamma = 70)$

Let us point out that the stability criterion derived above is a local criterion, which for the moment we have turned into a global one by the simplest prescription concerning the localization of the instability. However, Esser and Grossmann [14], on the basis of a detailed analysis of the basic destabilizing mechanisms, have recently shown that the stability boundary can be much better approximated by an analytic expression when being more precise on the instability localization. As a result, they found a stability criterion of the form $Ta = Ta_c$ provided one redefines the Taylor number as [15]:

$$Ta = -Re^2 \left(R_\Omega + 1\right)\left(R_\Omega + 1 - \frac{1}{\eta\xi^2}\right)\left(\frac{2\eta(\xi(\eta) - 1)}{(1 - \eta)}\right)^4$$

with

$$\xi(\eta) = r_p/r_i = 1 + \frac{1 - \eta}{2\eta}\Delta\left(a(\eta)\frac{d_n}{d}\right),$$

$$\frac{d_n}{d} = \frac{\eta}{1 - \eta} \left(\frac{1}{\sqrt{\eta(R_\Omega + 1)}} - 1 \right),$$

$$a(\eta) = (1 - \eta) \left(\sqrt{\frac{(1 + \eta)^3}{2(1 + 3\eta)}} - \eta \right)^{-1}, \tag{13.10}$$

where $\Delta(y)$ is a function equal to y if $y < 1$ and equal to 1 if $y > 1$. Re, the Reynolds number built on the mean shear $\bar{S}$, and R_Ω, the rotation number built on the mean angular velocity $\bar{\Omega}$, are defined as

$$Re = \frac{\bar{S} d^2}{\nu} = \frac{2}{1 + \eta} r_i d \frac{|\omega_o - \omega_i|}{\nu}, \tag{13.11}$$

$$R_\Omega = \frac{2\bar{\Omega}}{\bar{S}} = \frac{1 - \eta}{\eta} \frac{\eta \omega_i + \omega_o}{\omega_o - \omega_i}. \tag{13.12}$$

These control parameters, which at first sight may look artificial, are indeed based on dynamical propreties of the flow and therefore can be applied to any rotating shear flow. As a matter of fact, such parameters have already been used in rotating channel flows [16, 17]. They will be used when extending the analogy to the turbulent flow.

13.3 Weakly Nonlinear Analysis and Secondary Modes

Both Rayleigh–Bénard rolls and Taylor vortices dynamics can be represented by a field $\Psi = A(t, z)e^{iq_c x}$ where the amplitude $A(t, z)$ satisfies, near the onset of instability, the Ginzburg–Landau equation:

$$\tau_0 \frac{\partial A}{\partial t} = \epsilon A + \xi_0^2 \frac{\partial^2 A}{\partial z^2} - g|A|^2 A, \tag{13.13}$$

where $\epsilon = (Ta - Ta_c)/Ta_c$ or $(Ra - Ra_c)/Ra_c$ is the relative distance from the onset of instability. The real coefficients τ_0 and ξ_0 depend only on the linear terms in the flow equations and represent, respectively, the characteristic time and the coherence length of perturbations, g is the Landau saturation constant. The Ginzburg–Landau equation depends on the boundary conditions which differ from one system to another. Its validity for Rayleigh–Bénard convection has been confirmed in experiments by Wesfreid et al. [18]. For the Taylor–Couette flow, it has been derived from Navier–Stokes equations by many authors [19, 20, 21, 22, 23]. In the Taylor–Couette system, the effect of boundary conditions have been investigated by Dominguez-Lerma et al. [24] and Ahlers [25]. They showed that for cylinders with finite length L, the critical Taylor number depends on the aspect-ratio $\Gamma = L/d$ as follows:

$$\frac{Ta_c(\Gamma)}{Ta_c(\infty)} = 1 + \frac{2\pi^2 \xi_0^2}{\Gamma^2} + \xi_0^2 (q - q_c)^2, \tag{13.14}$$

where q_c and $Ta_c(\infty)$ are calculated values for infinitely long cylinders. The spatial variation of the pattern for Rayleigh–Bénard or Taylor–Couette flow can be represented by the real amplitude $A_0(t, z)$ and the phase ϕ as follows:

$$A(t, z) = A_0(t, z)e^{i\phi(t,z)} \tag{13.15}$$

The real amplitude $A_0(t, z)$ satisfies the Ginzburg–Landau equation (13.13) and the phase satisfies the following equation which is the simplest phase equation valid only for $\epsilon << 1$:

$$\frac{\partial \phi}{\partial t} = D\frac{\partial^2 \phi}{\partial z^2} \quad \text{with} \quad D = D_0\frac{\epsilon - 3\xi_0^2(q - q_c)^2}{\epsilon - \xi_0^2(q - q_c)^2} \tag{13.16}$$

The phase dynamics of a stationary pattern was investigated theoretically by Pomeau and Manneville [26] who established the phase equation from general principle. When the diffusion coefficient vanishes at the Eckhaus boundary $\epsilon = 3\xi_0^2(q - q_c)^2$, the stationary pattern pertains a modulational instability called Eckhaus instability. The experimental investigation of the phase dynamics in Rayleigh–Bénard convection was performed by Wesfreid and Croquette [27] who found $D_0 = 2.35$ for a silicon oil with $Pr = 490$. In the Taylor–Couette flow, the phase dynamics was investigated theoretically by Paap and Riecke [28] who have delimited the Eckhaus instability domains. Experiments were performed by different groups [25, 29, 30, 31]. The latter found the value of the phase coefficient for Taylor vortex flow to be about $D_0 = 1.643$ for a system with a radius ratio $\eta = 0.883$ and an aspect-ratio $\Gamma = 70$.

Note that in the case of sufficiently counterrotating cylinders, the first instability is a supercritical Hopf bifurcation that gives rise to a spiral vortex [32, 33], whose weakly nonlinear behavior can be described by the complex Ginzburg–Landau equation [22, 23, 34, 35]. Considering the analogy between the two systems when dealing with the secondary instabilities, there are still important similarities. In both cases, the transition to turbulence occurs via a cascade of successive bifurcations in a globally supercritical scenario. Also, the visual inspection of secondary flows calls for a formal analogy. Figure 13.3 displays modulated patterns obtained in both systems. They look strikingly similar. For details on the observed flows, one may refer to [36] for Rayleigh–Bénard convection and to [32] and [33] for the Taylor–Couette system. Unfortunately, no systematic study of the analogy has been proposed.

13.4 Turbulent Regime

So far, we only considered the analogy in the regimes where perturbations are small, resulting in linear or weakly nonlinear equations of motions. Dubrulle and Hersant [37] have shown that this analogy can be extended into the turbulent regime. Indeed, at small scales, turbulent motions are mainly slaved to the large scales. Their dynamics can then be approximated by a linearized equation of motions, in the spirit of the equations of the Rapid Distortion Theory. The

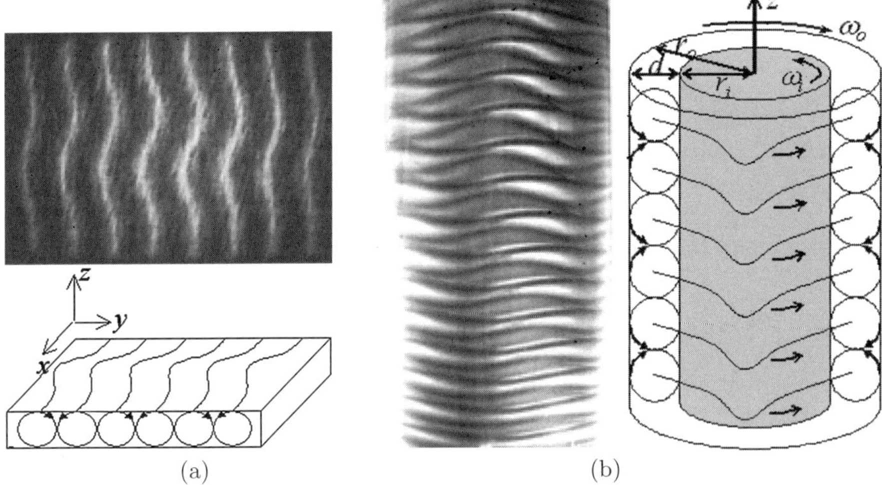

Fig. 13.3. Pictures and schematic views of the secondary flows in (a) the Rayleigh–Bénard convection (courtesy of F. Daviaud from SPEC-CEA, Saclay) and (b) the Taylor–Couette flow.

same remark can be applied to Rayleigh–Bénard convection, because an effective shear is generated by large-scale motions. This extension of the analogy into the turbulent regime yields a number of interesting consequences on the turbulent transport and profiles.

13.4.1 Turbulent Transport

A classical quantity in convection is the nondimensional heat transfer $Nu = Hd/\kappa\Delta T$, called the Nusselt number, where H is the heat transfer. The analogue of this in the Taylor–Couette flow is a nondimensional angular momentum transfer (with respect to the laminar value). This can be written using the nondimensional torque $G = T/\rho\nu^2 L$, where L is the cylinder length and T the torque:

$$Nu_* \equiv \frac{G}{G_{laminar}} = \frac{G}{Re}\frac{(1-\eta)^2}{2\pi\eta} \qquad (13.17)$$

The normalization by $G_{laminar}$ ensures that in the laminar case, $Nu_* = 1$, as in the convective analogue.

Near threshold, theoretical [38] and experimental [39] studies of convection lead to identification of two regimes just above the critical Rayleigh number. For $\epsilon = (Ra - Ra_c)/Ra_c \leq 1$, one has a linear regime in which

$$(Nu - 1)\frac{Ra}{Ra_c} = K_1\epsilon, \qquad (13.18)$$

where the constant K_1 depends on the Prandtl number. For $Pr = 1$, it is $K_1 \approx 1/0.7 = 1.43$ [38]. For larger ϵ, a scaling regime appears in which [39]

$$(Nu - 1)\,\frac{Ra}{Ra_c} = K_2 \epsilon^{1.23}, \tag{13.19}$$

where K_2 is a constant that is not predicted by the theory.

In the Taylor–Couette flow, the analogy yields the same scaling laws with Ta and Nu_* instead of Ra and Nu, respectively. Figure 13.4 shows how the results of Wendt obtained for $\eta = 0.935$ near the instability threshold compare with these two predictions. One sees that the linear regime is indeed obtained for $\epsilon \leq 10$, and the scaling regime is obtained for larger values of $10 < \epsilon < 100$.

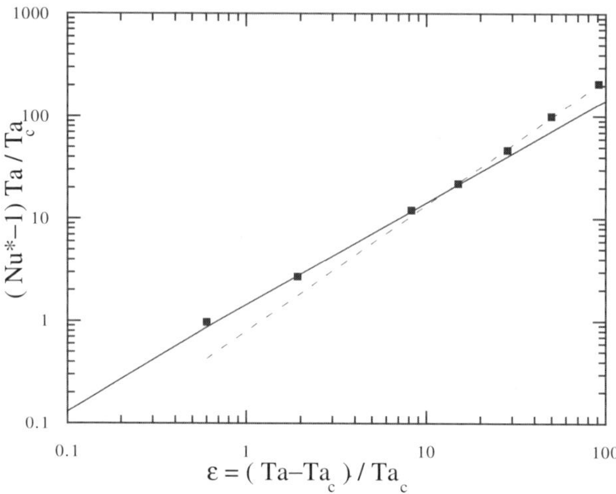

Fig. 13.4. Comparison of the theoretical near-instability onset behavior with Wendt's data [43]. The symbols are the experimental measurements. The two lines are the theoretical formula predicted by analogy with convection for $\epsilon < 1$ ($(Nu_* - 1)Ta/Ta_c = 1.43\epsilon$ and for $\epsilon > 1$ ($(Nu_* - 1)Ta/Ta_c \sim \epsilon^{1.23}$). In the latter case, the proportionality constant is not constrained by the analogy, and needs to be adjusted for a best fit.

Further from the threshold, one needs to compare the turbulent theories of convection. In the Taylor–Couette flow, transport properties in the turbulent regime have been investigated in a number of works. In the case of a rotating inner cylinder and the outer one at rest, one observes a tendency of the torque to behave as some power of the Reynolds number $G \sim Re^{\alpha}$ [40]. There is no clear consensus about this dependence yet: the marginal stability computations of King et al. [41] or Barcilon and Brindley [42] predict that the nondimensional torque should vary like $G \sim Re^{5/3}$. Old experimental data indicated the existence of two scaling regimes, one for $Re > 10^4$ where the exponent is 1.5, and another for a larger Reynolds number, where the exponent switches towards $1.7 - 1.8$ [43, 44, 45]. Recent high-precision experimental data yielded no region of constant exponent, and measured a "local" exponent $d\ln(G)/d\ln(Re)$ that varies continuously from 1.2 to 1.9, with a transition at $Re \sim 1.5 \times 10^4$ (for

$\eta = 0.7246$). This transition was later found to correspond to a modification of coherent structures in the flow [46].

This situation is reminiscent of turbulent convection where different scalings regime for Nu as a function of Ra are observed. In the classical theory of convection, one usually considers three regimes: a first one, labeled "soft turbulence", in which $Nu \sim Ra^{1/3}$ [47]; this regime is followed by a "hard turbulence" regime in which $Nu \sim Ra^{2/7}$ [48]; finally at very large Rayleigh numbers an ultra-hard turbulence regime occurs in which $Nu \sim Ra^{1/2}$ [49, 50]. More recently, Dubrulle [51, 52] used a quasi-linear turbulence model to predict slightly different dependence: at a low Rayleigh number, the dissipation is dominated by the mean flow, and $Nu = K_1 Ra^{1/4} Pr^{-1/12}$; at a larger Rayleigh number, the kinetic energy dissipation starts being dominated by velocity fluctuations, and the heat transport becomes $Nu Pr^{1/9} = K_2 Ra^{1/3} / \ln(Ra Pr^{2/3}/20)^{2/3}$. Finally at a very large Rayleigh number, the heat dissipation also becomes dominated by (heat) fluctuations, and $Nu = K_3 Ra^{1/2} / \ln(Ra/Ra_c)^{3/2}$. This classification summarized in Table 13.2 shows that it is unlikely that the three regimes co-exist in Taylor–Couette flow. Indeed, since the temperature analogue is related to the velocity, it might be impossible to excite velocity fluctuations without exciting pseudo-temperature fluctuations. This would mean a direct transition from regime 1 (mean flow dominated) to regime 3 (fluctuation dominated).

Table 13.2. Dissipation Mechanism in Both Flows when Increasing Re.

		$Re \nearrow \quad \longrightarrow$	
R.–B. dissipation	Mean flow	Velocity fluctuations	Velocity and heat fluctuations
T.–C. dissipation	Mean flow	Velocity fluctuations	

The theory predicts two regimes for the torque behavior:

- A "low Reynolds number" regime, where mean flow (Taylor vortices) dominates, in which

$$G = K_4 \frac{(3+\eta)^{1/4}(\eta Re)^{3/2}}{(1-\eta)^{7/4}(1+\eta)^{1/2}}. \qquad (13.20)$$

- A second regime in which

$$G = K_7 \frac{(3+\eta)^{1/2}}{(1-\eta)^{3/2}(1+\eta)} \frac{(\eta Re)^2}{\left(\ln[(K(\eta)(\eta Re)^2]\right)^{3/2}}, \qquad (13.21)$$

where

$$K(\eta) = K_8 \frac{(1-\eta)(3+\eta)}{(1+\eta)^2}. \qquad (13.22)$$

The value of the unknown coefficients is found by analogy or by best fit of the data. The analogy with convection predicts that $K_4 = 2\pi K_1$. The small aspect-ratio convective experiment extrapolated at large aspect-ratio gives $K_1 =$

0.75×0.31, which translates into $K_4 = 1.46$. This is in very good agreement with the prefactor measured by Wendt [43] who found the same exact dependence in η and Re for $400 < Re < 10^4$, and with a prefactor of $K_4 = 1.45$. The other coefficients have to be fitted (because they occur in a situation where regime 2 is bypassed). Using the data of Swinney and Lewis at $\eta = 0.72$, one can obtain $K_7 = 0.33$ and $K_8 = 10^{-4}$. Note that the analogy enables a prediction of the dependence of the torque as a function of other geometrical parameters, such as the gap. This provides a parameter-free prediction, to be tested by experiment with different gap. The result is shown in Figure 13.5, for three different gap values $\eta = 0.68, 0.85, 0.935$. The agreement is excellent.

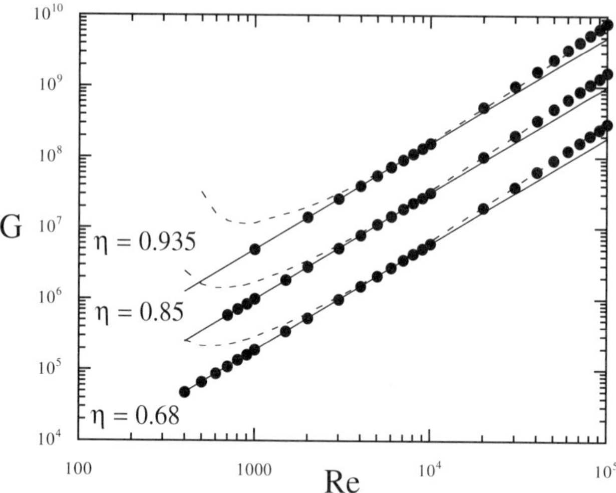

Fig. 13.5. Torque versus Reynolds number in Taylor–Couette flow for different gap widths $\eta = 0.68$, $\eta = 0.85$, and $\eta = 0.935$. The symbols are the data of [43]. The lines are the theoretical formula obtained in the soft and ultra-hard turbulence regimes and computed using the analogy with convection. Soft turbulence Equation (13.20) (full line); ultra-hard turbulence Equation (13.21) (dotted line). There is no adjustable parameter in this comparison, all the constants being fixed either by the analogy with convection, or by the comparison with the data of [46].

13.4.2 Velocity Profile

Mean profiles have been measured recently for different Reynolds numbers by Lewis and Swinney [46] in the case with the outer cylinder at rest. They observed that the mean angular momentum $\bar{\mathcal{L}} = r < v_\theta >$ is approximately constant within the core of the flow: $\bar{\mathcal{L}} \sim 0.5 r_i^2 \omega_i$ for Reynolds numbers between 1.4×10^4 and 6×10^5. At low Reynolds numbers, this feature can be explained by noting that reducing the angular momentum is a way to damp the linear instability, and, thus, to saturate turbulence. This is in good agreement with turbulent convection, which reaches saturation through the constancy of the temperature within the core of the flow. However, at larger Reynolds numbers, the angular momentum ceases to be constant, and instead reaches another regime, with constant shear approximately equal to 1/4 of the laminar value.

This is in agreement with a theoretical prediction by Busse [53]. He claims that the mean profiles obtained by Lewis and Swinney are actually in good agreement with a profile obtained by maximizing turbulent transport in the limit of high Reynolds numbers. This profile is given by an azimuthal circulation [54], as follows:

$$u^\infty(r) = \frac{\omega_i - \omega_o}{4r} \frac{\eta^2}{1 - \eta^2} + \frac{\omega_i \eta^2 (1 - 2\eta^2) + \omega_o (2 - \eta^2)}{2(1 - \eta^2)} \qquad (13.23)$$

In contrast, the same technique applied to convection predicts isothermal temperature profiles at large Rayleigh numbers, even though the maximizing principle differs only slightly from the maximizing Taylor–Couette flow. According to Busse, the difference originates in different ratios between the dissipation of the mean profiles and the fluctuating components of the extremelizing vector fields. Although this ratio approaches unity in the case of the extremilizing solution for the Couette problem, it reaches twice this value in the asymptotic case of convection. This may be due to the fact that the real convection is tridimensional (and not bidimensional as required by the analogy), resulting in more degrees of freedom.

13.4.3 Fluctuations

The analogy perhaps provides more success when dealing with fluctuations. In [46], the azimuthal turbulent intensity $i_\theta = \sqrt{<v_\theta^2>}/V_\theta$ was measured at midgap with hot film probes, where V_θ is the average value of the azimuthal velocity component of the turbulent flow. For $Re > 1 \times 10^4$, a best fit yields

$$i_\theta = 0.10 Re^{-0.125}. \qquad (13.24)$$

Using the analogy, this intensity is related to the temperature fluctuations at mid-gap, in the ultra-hard turbulent regime (regime 3). The total analogue temperature fluctuation in fact also includes vertical velocity fluctuations (Table 2). In an axisymmetric turbulence, one could therefore expect that the turbulent intensity measured by Lewis and Swinney would be proportional to the temperature analogue. Recent measurements of this quantity at Rayleigh numbers up to $Ra = 10^{15}$ have been performed by [55] in a low aspect-ratio helium experiment. They found $\theta/\Delta T = 0.37 Ra^{-0.145}$, but this was obtained in a regime where the Nusselt number varies as in regime 2 (velocity fluctuation dominates but not temperature fluctuation). Using the analogy, this would translate into a regime where $i_\theta \sim Re^{-0.29}$, in clear contradiction to the data of Lewis and Swinney (Figure 13.6). This might therefore be another proof of the absence of regime 2 in the Taylor–Couette flow.

238   Arnaud Prigent et al.

Unfortunately, we are not aware of laboratory temperature measurements in convective turbulence in the ultra-hard regime. In a previous analysis of temperature fluctuations in the atmospheric boundary layer (a large Rayleigh number medium, presumably being in the ultra-hard convective regime), Deardoff and Willis [56] showed that temperature fluctuations follow the free convection regime

$$\frac{\theta}{\Delta T} \propto \frac{Nu}{(PrRaNu)^{1/3}},\qquad(13.25)$$

where the proportionality constant is of the order 1. Using the analogy, this can be translated into

$$i_\theta = \frac{K_{11}}{\ln(Re/Re_c)}.\qquad(13.26)$$

Figure 13.6 shows the application of this scaling to the data of Lewis and Swinney. The best agreement with the experimental fit of Lewis and Swinney is obtained for a prefactor $K_{11} = 0.18$ and $Re_c = 44.4$.

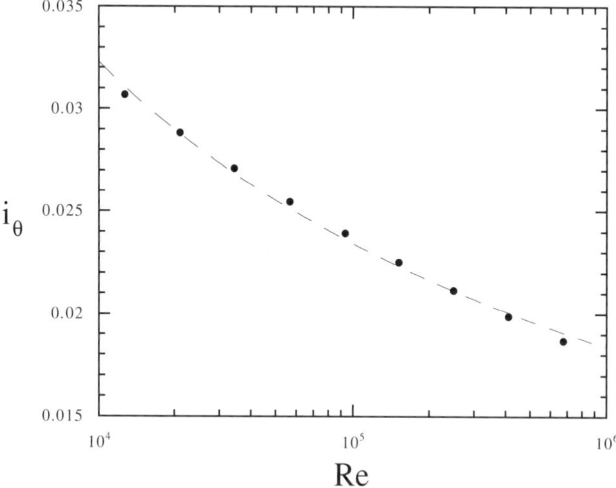

Fig. 13.6. Azimuthal velocity fluctuations in Taylor–Couette flow. The circles are the power-law fits of experimental measurements by [46]. The triangles are the power-law fit to the temperature fluctuations (analogue of azimuthal velocities) in helium by [55]. The line is the prediction obtained with the analogue of the free-convective regime [56].

13.5 Conclusion

We have reviewed in detail the analogy between the Taylor–Couette flow and Rayleigh–Bénard convection. Despite different symmetry properties, because in Rayleigh–Bénard convection both transverse directions are equivalent whereas in the Taylor–Couette system the axial and azimuthal directions are not equivalent, the instability mechanism appears to be the same in both systems. It is based on density stratification in the former and on angular momentum stratification in the second. The linear stability of the circular Couette flow in the small gap limit and that of Rayleigh–Bénard convection in the Boussinesq approximation lead to a similar perturbation equation, and thereby to the same threshold and critical wavenumber. The patterns appearing at onset of the primary and secondary instabilities also look similar. This analogy has been used to develop stability criteria in other systems, such as instabilities in liquid crystals, and Coriolis force induced instabilities [9, 57, 12, 58, 59]. Above, but not too far from the instability threshold, both Taylor vortices and Bénard rolls may be described in the framework of Ginzburg–Landau model. The phases obey the same equation and patterns may become unstable against Eckhaus instability. Further above the threshold, the secondary instabilities still present strong similarities. However, the weakly nonlinear behavior of the patterns and their secondary instability modes have subtle differences that call for careful investigation.

Finally, we have discussed the extension of the analogy to the turbulent regime when the mean shear, generated by large scales, is taken into account in the Rayleigh–Bénard system. Then, analogy allows theoretical predictions for the angular momentum transfer and velocity profiles. In turbulent convection, three regimes with different modes of dissipation are identified. At low Ra, the dissipation is dominated by mean flow and it is dominated by velocity fluctuations for larger Ra, and by velocity fluctuations at very large Ra. The analogy leads to the identification of two flow regimes in the Taylor–Couette flow corresponding to the first and the third ones in convection.

Acknowledgments

This work results from the collaboration program between Le Havre University and CEA in Saclay. We would like to thank F. Daviaud for his permanent support and P. Le Gal for interesting discussions on this work.

References

1. M. Couette, "Sur un nouvel appareil pour l'étude du frottement des fluides", *Ann. Chim. Phys.* (6) **21**, 433–510, 70 (1890).
2. A. Mallock, "Determination of the viscosity of water", *Phil. Trans. Roy. Soc. A* **187**, 41–56 (1896).
3. L. Rayleigh, "On the dynamics of revolving fluids", *Proc. R. Soc. London*, Ser. A **93**, 148–54 (1916).
4. L. Rayleigh, "On convective currents in a horizontal layer of fluid, when the higher temperature is on the under side", *Phil. Mag* (6) **32**, 529–46 (1916).
5. G.I. Taylor, "Stability of a viscous liquid contained between two rotating cylinders", *Phil. Trans. Roy. Soc. London*, Ser. A **223**, 289–343 (1923).
6. H. Jeffreys, "Some cases of instability in fluid motion", *Proc. R. Soc. London*, Ser. A **118**, 195–208 (1928).

7. S. Chandrasekhar, *Hydrodynamic and Hydromagnetic Stability*, Clarendon Press, Oxford (1961).
8. F.H. Busse, "Bounds for turbulent shear flow", *J. Fluid Mech.* **41**, 219 (1970).
9. E. Guyon, J.P. Hulin, and L. Petit, *Hydrodynamique Physique*, Savoirs actuels, Paris (2002).
10. E.L. Koscmieder, *Bénard Cells and Taylor Vortices*, Cambridge University Press, Cambridge (1993).
11. B.J. Bayly, "Three-dimensional centrifugal-type instabilities in inviscid two-dimensional flows", *Phys. Fluids A* **31**(1), 56–64 (1988).
12. I. Mutabazi, C. Normand, and J.E. Wesfreid, "Gap size effects on centrifugally and rotaionally driven instabilities", *Phys. Fluids A* **4**, 1199–1205 (1992).
13. P. Manneville, *Structures Dissipatives, Chaos et Turbulence*, Aléa Saclay, Saclay (1991).
14. A. Esser and S. Grossmann, "Analytic expression for Taylor–Couette stability boundary", *Phys. Fluids* **8**, 1814 (1996).
15. B. Dubrulle, O. Dauchot, F. Daviaud, F. Hersant, P.-Y. Longaretti, D. Richard, and J-P. Zahn,"Stability and turbulent transport in rotating shear flows: prescription from analysis of cylindrical and plane Couette flows", submitted to *Phys. Fluids* (2004).
16. D.K. Lezius and J.P. Johnston, "Roll-cell instabilities in rotating laminar and turbulent channel flows", *J. Fluid Mech.* **77**, 573 (1976).
17. O.J.E. Matsson and P.H. Alfredsson, "Curvature- and rotation-induced instabilities in channel flow", *J. Fluid Mech.* **210**, 537–563 (1990).
18. J.E. Wesfreid, Y. Pomeau, M. Dubois, C. Normand, and P. Bergé, "Critical effects in Rayleigh–Bénard convection", *J. Physique (Paris)* **39**, 725 (1978).
19. H. Yahata, "Slowly-varying amplitude of the Taylor vortices near the instability point", *Prog. Theor. Phys.* **57**, 347 (1977).
20. P. Tabeling, "Dynamics of the phase variable in the Taylor vortex system", *J. Phys. Lett.* **44**, 16 (1983).
21. P. Hall, "Evolution equations for Taylor vortices in the small-gap limit", *Phys. Rev. A* **29**, 2921 (1984).
22. Y. Demay and G. Iooss, "Calcul des solutions bifurquées pour le problème de Couette–Taylor avec les 2 cylindres en rotation", *J. Méca. Théor. Appl.*, numéro spécial, 193 (1984).
23. Y. Demay, G. Iooss, and P. Laure, "Wave patterns in the small gap Couette–Taylor problem", *Eur. J. Mech. B* **11**, 621 (1992).
24. M.A. Dominguez-Lerma, G. Ahlers, and D. S. Cannell, "Marginal stability curve and linear growth rate for rotating Couette–Taylor flow and Rayleigh–Bénard convection", *Phys. Fluids* **27**, 856 (1984).
25. G. Ahlers, "Experiments on bifurcation and one-dimensional patterns in nonlinear systems far from equilibrium", in *Lectures in the Sciences of Complexity*, ed. by D.L. Stein, Addison-Wesley, Redwood City, CA (1989).
26. Y. Pomeau and P. Manneville, "Stability and fluctuations of a spatially periodic convective flow", *J. Phys. Lett.* **40**, 610 (1979).
27. J.E. Wesfreid and V. Croquette, "Forced Phase Diffusion in Rayleigh–Bénard Convection ", *Phys. Rev. Lett.* **45**, 6340 (1980).
28. H. Paap and H. Riecke, "Drifting vortices in ramped Taylor vortex flow. Quantitative results from phase equation", *Phys. Fluids A* **3**, 1519 (1991).
29. M. Wu and C.D. Andereck, "Phase modulation of Taylor vortex flow ", *Phys. Rev. A* **43**, 2074 (1991).

30. M. Wu and C.D. Andereck, "Phase dynamics of wavy vortex flow", *Phys. Rev. Lett.* **67**, 1258 (1991).

31. M. Wu and C.D. Andereck, "Phase dynamics in the Taylor–Couette system", *Phys. Fluids A* **4**, 2432 (1992).

32. D. Coles, "Transition in circular Couette flow", *J. Fluid Mech.* **21**, 385–425 (1965).

33. C.D. Andereck, S.S. Liu, and H.L. Swinney, "Flow regimes in a circular Couette system with independently rotating cylinders", *J. Fluid Mech.* **164**, 155–183 (1986).

34. R. Tagg, W.S. Edwards, and H.L. Swinney, "Nonlinear standing waves in Couette-Taylor flow", *Phys. Rev. A* **39**, 3734 (1989).

35. R. Tagg, "The Couette-Taylor problem", *Nonlinear Science Today* **4**, 1 (1994).

36. F.H. Busse, "Transition to turbulence in Rayleigh–Bénard convection", in *Hydrodynamic Instabilities and the Transition to Turbulence* eds. H.L. Swinney and J.P. Gollub, Springer, New York (1981).

37. B. Dubrulle and F. Hersant, "Momentum transport and torque scaling in Taylor–Couette flow from an analogy with turbulent convection", *Eur. Phys. J. B* **26**, 379 (2002).

38. A. Schlüter, D. Lortz, and F. Busse, "Transition to turbulence in Rayleigh–Bénard convection", *J. Fluid Mech.* **23**, 129 (1965).

39. J.K. Platten and J.C. Legros, *Convection in liquids*, Springer, New York (1984).

40. D.P. Lathrop, J. Fineberg, and H.L. Swinney, "Transition to shear-driven turbulence in Couette–Taylor flow", Phys. Rev A **46**, 6390 (1992).

41. G.P. King, Y. Li, W. Lee, H.L. Swinney, and P.S. Marcus, "Wave speeds in wavy Taylor vortex flow", *J. Fluid Mech.* **41**, 365 (1984).

42. A. Barcilon and J. Brindley, "Organized structures in turbulent Taylor–Couette flows at a very high Taylor number", *J. Fluid Mech.* **143**, 429 (1984).

43. F. Wendt, "Turbulente Stromungen zwischen zwei rotierenden konaxialen Zylindern", *Ingenieur-Archiv.* **4**, 577 (1933).

44. G.I. Taylor, "Fluid friction between rotating cylinders I. - Torque measurements", *Proc. R. Soc. London A* **157**, 546 (1936).

45. P. Tong, W.I. Goldburg, J.S. Huang, and T.A. Witten, "Anisotropy in turbulent drag reduction", *Phys. Rev. Lett.* **65**, 2780 (1990).

46. G.S. Lewis and H.L. Swinney, "Velocity structure functions, scaling, and transitions in high-Reynolds-number Couette–Taylor flow", *Phys Rev. E* **59**, 5457 (1999).

47. F. Heslot, B. Castaing, and A. Libchaber, "Transitions to turbulence in helium gas ", *Phys. Rev. A* **36**, 5870 (1987).

48. B. Castaing, G. Gunaratne, F. Heslot, L. Kadanoff, A. Libchaber, S. Thomae, X-Z. Wu, S. Zaleski, and G. Zanetti, "Scaling of hard thermal turbulence in Rayleigh–Bénard convection", *J. Fluid Mech.* **204**, 1 (1989).

49. R. Kraichnan, "Mixing-length analysis of turbulent thermal convection at arbitrary Prandtl number", *Phys. Fluids* **5**, 1374 (1962).

50. X. Chavanne, F. Chilla, B. Castaing, B. Hébral, B. Chabaud, and J. Chaussy, "Observation of the ultimate regime in Rayleigh–Bénard convection", *Phys. Rev. Lett.* **79**, 3648 (1997).

51. B. Dubrulle, "Scaling laws predictions from a solvable model of turbulent thermal convection", *Europhys. Letters* **51**, 513 (2000).

52. B. Dubrulle, "Logarithmic corrections to scaling in turbulent thermal convection", *Euro. Phys. J. B* **21**, 295 (2001).

53. F.H. Busse, "Transition to turbulence in Rayleigh–Bénard convection", in *Nonlinear Physics of Complex Systems* ed. G. Parisi, S.C. Muller, and W. Zimmermann, *Lecture Notes in Physics* **476**, 1, Springer, New York (1996).

54. F.H. Busse, "Transition to turbulence in Rayleigh–Bénard convection", Arch. Rat. *Mech. Anals* **47**, 28 (1972).

55. J.J. Niemela, L. Skrbek, K.R. Sreenivasan, and R.J. Donnelly, "Turbulent convection at very high Rayleigh number", *Nature* **404**, 837 (2000).

56. J.W. Deardoff and G.E Willis, "Investigation of turbulent thermal convection between horizontal plates ", *J. Fluid Mech.* **28**, 675 (1967).

57. E. Guyon and P. Pieranski, "Convective instabilities in nematic liquid crystals ", *Physica* **73**, 184 (1974).

58. I.H. Herron, "Stability criteria for flow along a convex wall", *Phys. Fluids A* **3**, 1825 (1991).

59. D.J. Tritton and P.A. Davies, in *Hydrodynamic Instabilities and the Transition to Turbulence*, H.L. Swinney and J.P. Gollub (eds.), Springer, New York (1981).

Index

Springer Tracts in Modern Physics